T0202288

21

TESI

THESES

tesi di perfezionamento in Matematica per le Tecnologie Industriali sostenuta l'8 ottobre 2015

COMMISSIONE GIUDICATRICE
Luigi Ambrosio, Presidente
Giacomo Bormetti
Franco Flandoli
Fabrio Lillo
Augusto Neri
Davide Erminio Pirino

Andrea Bevilacqua
Scuola Normale Superiore
Piazza dei Cavalieri, 7
56126 Pisa
e
Istituto Nazionale di Geofisica e Vulcanologia
Via Uguccione della Faggiola, 32
56126 Pisa

Doubly Stochastic Models for Volcanic Hazard Assessment at Campi Flegrei Caldera

Andrea Bevilacqua

Doubly Stochastic Models for Volcanic Hazard Assessment at Campi Flegrei Caldera

EDIZIONI
DELLA
NORMALE

© 2016 Scuola Normale Superiore Pisa

ISBN 978-88-7642-556-1
e-ISBN 978-88-7642-577-6

Contents

Acknowledgements

This thesis was developed and supported thanks to the agreement between Istituto Nazionale di Geofisica e Vulcanologia and Scuola Normale Superiore di Pisa (Convenzione INGV-SNS 2009-2011). The thesis has been partially developed in the framework of the project 'V1 - Stima della pericolosità vulcanica in termini probabilistici' funded by Dipartimento della Protezione Civile. Partial support was also provided by the EU-funded project 'MEDSUV - Mediterranean Supervolcanoes' (grant 308665) and the COST Action 'EJN - Expert Judgement Network: bridging the gap between scientific uncertainty and evidence-based decision making' (IS1304). The contribution and support of ideas of many colleagues participating to the above projects are acknowledged. Some of the results of this study have been published on the Journal of Geophysical Research - Solid Earth, in the two research articles [17] and [118]. Furthermore some preliminary results have been used for producing the scientific report [117] for Dipartimento della Protezione Civile and Commissione Nazionale Grandi Rischi. However this thesis does not necessarily represent official views and policies of Dipartimento della Protezione Civile.

Chapter 1
Introduction

1.1. General synopsis

Volcanic eruptions are the surface discharges of gas and magma (*e.g.*
[129, 30]), which is a mixture of molten rock, suspended crystals and
dissolved gas; it sometimes includes also gas bubbles and rock fragments
(see Figure 1.1). Magma has complex properties that reflect the changing
proportions of its components and chemistry, and it is capable of intrusion
into adjacent rocks forming dikes and sills, extrusion onto the surface as
lava, and explosive ejection as tephra to form pyroclastic rock (see [73,
83, 84]). The widely accepted qualitative model for a volcanic system
assumes the presence of one or more magma reservoirs below the surface
(*e.g.* [103]), that may become over-pressurized because of the injection
of new high temperature magma or because of a structural weakening of
the surrounding rock, hence overcoming the critical pressure required for
the propagation of magma to the surface, opening a vent (or fissure) and
erupting (*e.g.* [146, 102, 149]). A volcano is any geographical feature
built by volcanic eruptions (*e.g.* [144, 52]).

The hazards are all the processes that produce danger to human life and
infrastructures while the risk is the potential or possibility that something
bad will happen because of the hazards. Volcanic eruptions can produce
huge risks, in terms of human losses, environmental consequences, and
economic costs: any effective volcanic risk mitigation strategy requires a
scientific assessment of the future evolution of the volcanic system and
its eruptive behaviour (*e.g.* [91, 158, 159]). Compared with other natural
hazards as earthquakes and severe weather phenomena, volcanic hazards
are characterized by long duration (weeks-months), large areas of impact
(national-international), different types of hazards during a single event,
recognizable precursors like ground deformation, seismicity, temperature
changes, anomalous discharge of gas and geothermal fluids (see [74]).

Volcanic hazards are numerous and showing diverse duration, force and
range: tephra fallout from umbrella clouds (*e.g.* [147, 29, 19, 157]) can
mantle vast areas with layers of pumice and ash, causing roof collapses,
the shutdown of road traffic and lifelines, destroying crops, damaging

Figure 1.1. Small explosive eruptions at Sakurajima volcano - July 19 (left) and 22 (right), 2013, Japan. Personal photos of A. Bevilacqua.

high voltage lines, and affecting human and animal health; pyroclastic flows, also called pyroclastic density currents (PDCs), which are fast-moving and highly destructive gravity currents of hot gas and rocks with up to more than 200 km/h speed (see [61, 116, 36, 23]), can raze and bury the most of the buildings and trees in their path, killing even sheltered people almost instantaneously and potentially generating large fires; lahars, which are mud flows composed of a slurry of pyroclastic material, rocky debris, and water forming from the mobilization of rapidly accumulated ash by rain or from the quick melting of volcanic glaciers (jökulhaups), can be as much devastating as PDCs (see [152, 112, 76]). Other volcanic hazards include lava flows capable of burning and burying roads, properties, trees and even buildings in their path and potentially killing people if the flow is fast enough (*e.g.* [82, 93]); large ballistics even weighting thousands of kilograms which can reach kilometers of distance from the erupting vent; toxic gasses and aerosols emissions which the volcano can rapidly release in huge amounts and which can then be transported even thousands of kilometers by the wind (see [156, 140]); fine ash pollution injected into the high atmosphere, which is capable of damaging critical components of airframes and may reach huge distances (*e.g.* [127, 78]); highly destructive tsunamis caused by partial or total collapse of the volcano edifice (even caused by underwater eruptions; see [155, 47]) and which may travel even across oceans. Such a variety of hazardous phenomena, the potentially global impact of volcanic eruptions, and the consideration that hundreds of millions of people in all continents live close enough to active volcanoes to be substantially affected by their activity, put volcanic risks among the most relevant natural risks on Earth.

Estimating the probability of an eruption event, its size, location, time and type is a very difficult issue because of the lack of detailed information on the deeper portions of the volcanic system and because of the high number of degrees of freedom, often nonlinearly coupled, that characterize the physical processes controlling it. Uncertainty is a key issue in volcanology because the exact form of future eruptions is not predictable: there is not an easy way to estimate the size of an event from the characteristics of its precursor symptoms, nor is there a simple rule to forecast a time interval between the onset and the climax of explosive activity (sometimes shorter than one day). In contrast to deterministic predictions, probabilistic approaches attempt to quantify the inherent uncertainties instead of trying to remove them, utilizing all the available information, but paying the cost of obtaining only probability distributions instead of precise forecasts (*e.g.* [40]). Probabilistic eruption assessment is currently the primary scientific basis for planning rational risk mitigation actions as well as for land use and emergency planning (*e.g.* [9, 11]).

It is important to distinguish between two classes of probability forecasts in volcanology: one contains the short-term assessments (*e.g.* [142]), which are typically of interest in managing evolving episodes of volcanic unrest and are mostly driven by the information provided by monitoring anomalies, like the occurrence of one or more signals outside a background range (see [125, 137]). The second class includes the long-term assessments (sometimes called background or base-rate), which are mostly required for land use and evacuation planning, but constitute also the necessary background for implementing a robust short-term probability model (see also [107]). They are primarily based on the available past eruption data and on the structural features of the system: such geological data can go back millennia and they typically come from distal tephra records, proximal volcanic products datation, and exposed faults or fractures measurements. Moreover, the geologic record often incompletely preserves evidence of smaller eruptions, and burial of older deposits is common. In general long-term assessments should reasonably be assumed more relevant than short-term assessments for producing probability forecasts on a longer time interval in the future, until the information on which they are based will considerably change. In this thesis we will focus on this second type of assessments.

Caldera volcanoes form during the largest and most powerful explosive eruptions: they are depressions left as a consequence of structural collapses following the ejection of colossal masses of magma in a short time (*e.g.* [21]). Between these large and infrequent events the calderas often present smaller explosive and dome-forming eruptions, possibly coming from shallow reservoirs and influenced by a very complex network of geologic structures under the surface, often associated with relevant hydrothermal systems. Vast areas around the calderas are covered by erupted ash and pumice, and the volcanoes themselves have the form of large craters. Probability hazard assessment is particularly complex for calderas due to the potentially sparse pattern of eruptions and the large variability of eruptions sizes and types. In addition, caldera volcanoes may persist in unrest conditions for decades, periodically showing precursor signals that would almost certainly lead to an eruption if observed at more typical central volcanoes. On the contrary, some observations show that calderas can originate a new eruption following a phase characterized by signals much less relevant than those observed in other periods not followed by any eruption. These reasons increase the importance of having a robust hazard model based on all the information about past behaviour of the volcano even during a crisis.

1.1.1. The Campi Flegrei caldera

Campi Flegrei is a volcanic caldera with a diameter of about 12 km and the town of Pozzuoli at its center (Figure 1.2; [128, 133, 122]). The northern and western parts of the caldera are above sea level and characterized by the presence of many dispersed cones and craters, whereas the southern part is principally submarine and extends into Golfo di Pozzuoli. Its name comes from the Greek $\varphi\lambda\epsilon\gamma\omega$, meaning 'to burn', indeed it is the most active caldera in Europe having had more than 70 eruptions within the last 15 ka (see [132, 57, 122, 97, 145]). Activity started more than 80 ka BP (*e.g.* [138, 153]) and includes the generation of the large caldera-collapse Campanian Ignimbrite eruption (CI, ~40 ka BP; [55, 71]) and the second major caldera-collapse eruption of the Neapolitan Yellow Tuff (NYT, ~15 ka BP; [121, 48]). In the last 15 ka, intense and mostly explosive volcanism and deformation has occurred within the NYT caldera, along its structural boundaries as well as along faults within (*e.g.* [57, 122, 97, 145]). Eruptions were closely spaced in time, over periods from a few centuries to a few millennia, with periods of quiescence lasting several millennia. As a consequence, activity has been generally subdivided into three distinct epochs, *i.e.* Epoch I, 15 - 10.6 ka; Epoch II, 9.6 - 9.1 ka, and Epoch III, 5.5 - 3.8 ka BP (*e.g.* [122, 145]). Simultaneous eruptions from different sectors of the caldera have also occurred at least during the Epoch III (see [97]). The most recent eruption was that of Monte Nuovo in 1538 AD, 477 years BP (*e.g.* [56, 58, 79]). Volcanism was also generally preceded by broadly distributed ground deformation phenomena leading to remarkable uplift of the central part of the caldera, *e.g.* larger than 100 meters in the last 10.5 ka (see [57, 97]) and several meters before the Monte Nuovo eruption (see [63, 79]).

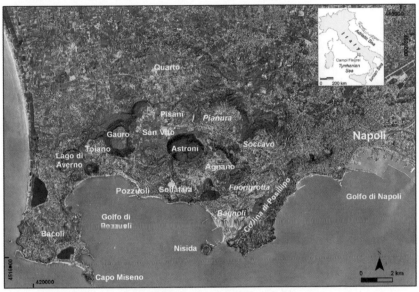

Figure 1.2. Mosaic of orthophotos of Campi Flegrei caldera and surrounding areas (including part of the city of Naples on the east) showing the large urbanization inside and around this active volcano, modified from [17].

In recent decades, Campi Flegrei has exhibited significant deformation phenomena in the central part of the caldera that produced a dome-like structure having a base diameter of about 6-7 km with an uplift of several tens of meters from the sea bottom, centered on the town of Pozzuoli (*e.g.* [16, 50]). For instance, in 1982-1984, there was rapid uplift of the center of the caldera of about 1.8 m. Since then, the caldera surface has been slowly subsiding, but punctuated by significant uplift episodes. Changes in the gas composition of fumaroles were measured in 2006 and again in 2011-2012 (see [34]). Based on the above information, and with more than three hundred thousand people living within the caldera, the volcanic risk at Campi Flegrei is considered to be substantial. Indeed Campi Flegrei is a densely populated and active caldera characterized by predominantly explosive eruptive activity (*e.g.* [132, 122, 145]). Key features of this activity have been eruptions from different vents scattered within the caldera, with individual events spanning a large range of eruptive scales. The products of the explosive activity can be found over most of the Campanian region in conspicuous pyroclastic deposits generated by tephra fallout and pyroclastic flows. PDCs represent the main hazard of this volcanic system (*e.g.* [132, 57, 122]): due to their velocity, temperature and particle concentrations, they can produce heavy damage to urban structures and lethal conditions for human beings (see [13, 116]). Given the very high urbanization of the caldera itself and its proximity to the city of Naples, it is of prime importance that areas which may potentially be affected by pyroclastic flows are identified and ranked in terms of exposure likelihood in order that civil authorities can prepare suitable mitigation measures (*e.g.* [14, 115]).

1.1.2. Doubly stochastic modelling

This thesis includes three main chapters, each one dedicated to a different part of a broad study aimed at the volcanic hazard probability assessment at Campi Flegrei caldera, with particular attention to the uncertainty quantification. A large effort is dedicated to the mathematical formalization of many geological assumptions and practical method implemented. We reached the objective of constructing:

- a map of probability for the location of the next eruptive vent (see also [17]);
- a probability distribution for the size of the next PDC, and a map of probability for the PDC invasion hazard (see also [118]);
- a time-space probability model for the time of the next eruption, focused on the vents clustering.

In particular, Campi Flegrei volcano is assumed as a complex random system that must be assessed with incomplete and uncertain information. In all the chapters we cope with problems affected by a very large uncertainty and for this reason the models constructed are, implicitly, doubly stochastic. The meaning of this assumption is that the location of the next eruptive vent or the size of the next PDC phenomenon (just to give the two main examples) cannot be easily forecast using a simple probability distribution. There are some sources of epistemic uncertainty even affecting the definition of the probability distribution itself, so we will follow the approach to assume a double structure below the sample space.

The basic idea of a doubly stochastic model is that an observed random variable is modelled in two steps: in one stage, the distribution of the observed outcome is represented using one or more parameters; at a second stage, some of these parameters are treated as being themselves random variables. This raises the definition of a double probability space with two different probability frameworks: one is expected to describe the physical variability of the system (sometimes called aleatoric uncertainty), the other assesses the epistemic uncertainty due to the imperfect knowledge of the system under study. This distinction is not always easy to do in practice because the two frameworks are correlated and convolved in producing the available observations, but it is very important because it corresponds to the distinction between an intrinsic randomness of the system and the additional uncertainty that affects its representation, originating from the incomplete information about past behaviour and expressing the degree of belief about alternative assumptions. In summary, the probability measures representing epistemic uncertainty and physical randomness are supported on two separated spaces, the first influencing the second (see Figure 1.3).

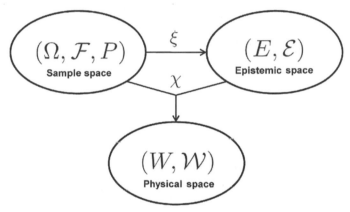

Figure 1.3. General scheme for a doubly stochastic model. The random variables ξ and χ respectively sample the epistemic assumptions and the physical observables (see Definition 1.1). With each measurable space we report also its σ-algebra; we include also the probability measure P on which the samples rely.

In this study the space $(W, \mathcal{W})$ could be thought as the space containing all the possible sequences of observables of the phenomena of interest, *i.e.* the future volcanic eruptions at Campi Flegrei. In particular we will assume it contains at least the spatial location of the next event (a variable X in $\mathbb{R}^2$, Chapter 2 and 3), the areal size invaded by the next PDC (a variable Y in $\mathbb{R}_+$, Chapter 3), the time-space pattern of the next eruptive activity (a point process Z, Chapter 4). Instead the space $(E, \mathcal{E})$ could be thought as a space of parameters which rule the model that we develop for representing the phenomena of interest: for example it will include the n-uples of possible responses to the expert judgement questions (Chapters 2 and 3) and the possible time sequences for the uncertain past eruption record (Chapter 4). The sample space $(\Omega, \mathcal{F}, P)$ has the motivation of putting together the physical and the epistemic spaces in a global framework. In general, all the random samples assessed are defined on it: for the purpose of numerical simulation it could be thought as containing all the seeds of the pseudo-random values generator inside the computer and it permits to develop a logical structure that is most natural to formalize all the Monte Carlo simulations that we implemented.

Definition 1.1 (The structure of uncertainty). Let ξ be a random variable from the sample space $(\Omega, \mathcal{F}, P)$ to the measurable space $(E, \mathcal{E})$ representing the epistemic variability of the sources of uncertainty considered, and let η be its probability law. Let χ be a random variable from $(E \times \Omega, \mathcal{E} \otimes \mathcal{F}, \eta \otimes P)$ to the measurable space $(W, \mathcal{W})$ representing the physical variability of the volcanic system for each occurrence of the sources of uncertainty. For each $e \in E$, we define on $(W, \mathcal{W})$ the image measure $M(e) = \chi(e, \cdot)_\sharp(P)$.

It is easy to see that M is a well defined random measure parameterized on E and supported on $(W, \mathcal{W})$, that represents the possible probability distributions of all the volcanic variables of interest, such as the next vent location or the size of the next PDC. Our purpose is to separately model the epistemic uncertainty affecting our knowledge and the physical variability of the problem. We will take into account several random variables on $(\Omega, \mathcal{F}, P)$ affected by a large uncertainty, but very important for hazard assessment: the random measure M permits us to define random versions of them, and we will rely on the probability η for convolving such versions and for assessing their uncertainty.

Various statistical approaches either non-parametric, semi-parametric or completely parametric will be adopted for the geologic data representation and the available observations implementation, and in addition, also the degree of belief of the scientific community about alternative assumptions will be quantified. Any procedure that permits the formalization of the opinion of a group of experts is called expert judgment (or elicitation) technique, and these are capable of evaluating even the epistemic uncertainties: heterogeneous groups of experts, performance based scores and structured procedures for combining the different responses will be aimed at decreasing the degree of subjectivity affecting the estimates. For these reason the adoption of such methods is of the main importance in this thesis ([41, 7, 68] and Chapter 5).

1.2. Vent opening probability maps

In this chapter we produce new background (sometimes also referred to as long-term or base-rate) probability maps of vent opening of the Campi Flegrei caldera by incorporating information from some of the most recent studies, specifically focusing on some of the key epistemic uncertainties of the volcanic system. In particular, the maps express the probability of vent opening conditional on the occurrence of a new eruption in the foreseeable future. This is done by considering the eruptive record of Campi Flegrei in the last 15 ka as well as the distribution of key structural features, such as faults and fractures, within the caldera. The probability model that we assumed is doubly stochastic, in the sense that the probability values representing the spatial physical variability affecting the vent opening process are themselves affected by epistemic uncertainty. The sources of epistemic uncertainty considered relate to the uncertain locations of past vents, the incompleteness of the eruptive record, and the uncertain weights given to the different volcanic system variables under consideration. We followed a structured elicitation with alternative pooling procedures, thus creating percentile maps associated with the sources of epistemic uncertainty considered, in addition to a map of mean probability. This product is of critical importance since it is the starting point for making probabilistic maps for the main hazardous phenomena that could be related to this caldera, including pyroclastic flows and ash fallout; it also provides, together with the collected monitoring data, the framework for mapping short-term vent openings.

1.2.1. Formal definition of the vent opening probability maps

The following is the general definition of the vent opening map whose explicit construction, estimating epistemic uncertainties and their influence on the physical variability, is the purpose of this chapter.

Definition 1.2 (The vent opening probability map). Let $A \subseteq \mathbb{R}^2$ be a domain representing the area of the volcanic system, considered with its Borel sigma algebra $\mathcal{B}(A)$. Let X be a random variable from the sample space $(\Omega, \mathcal{F}, P)$ to $(A, \mathcal{B}(A))$, representing the location of the next eruptive vent at Campi Flegrei. Let μ_X be the probability measure that is the law of X on $(A, \mathcal{B}(A))$; it is called a vent opening probability map.

We assess an estimation for the unknown measure μ_X and quantify its epistemic uncertainty defining a random measure. We follow the nontrivial probability structure of Definition 1.1, just projecting the space $(W, \mathcal{W}, M)$ on $(A, \mathcal{B}(A))$.

Definition 1.3 (Conditional vent opening probability map). Let π^1 be a measurable function from $(W, \mathcal{W}, M)$ to $(A, \mathcal{B}(A))$, representing the projection of the physical space onto the space of the vent opening location. We assume that

$$X(\omega) = \pi^1 \left(\chi(\xi(\omega), \omega) \right), \text{ for almost every } \omega \in \Omega,$$

and we define the random variable $\check{X}$ from $(E \times \Omega, \mathcal{E} \otimes \mathcal{F}, \eta \otimes P)$ to $(A, \mathcal{B}(A))$ as

$$\check{X}(e, \omega) := \pi^1 \left(\chi(e, \omega) \right).$$

For each $e \in E$ the random variable $\check{X}(e, \cdot)$ on $(\Omega, \mathcal{F}, P)$ represents the location of the next eruptive vent at Campi Flegrei once adopted the epistemic assumption e. Its law $\mu_{\check{X}}(e)$ is called vent opening probability map conditional on the epistemic assumption e.

In particular $\mu_{\check{X}} = \pi^1_\sharp(M)$: a well defined random measure on $(A, \mathcal{B}(A))$, parameterized on E.

1.2.2. Description of the achievements

The vent opening probability map is constructed as a weighted mixture of several probability distributions related to the different volcanological features available. Experts judgment outcomes indicate that past vent locations are the most informative factors governing the estimates of the probabilities of vent opening, followed by the locations of faults and fractures; also a uniform spatial density distribution for vent opening over the whole caldera is implemented in order to account for a possible lack of correlation with the variables considered. In addition to the mean probability map $E^E[\mu_{\check{X}}(\cdot)]$ representative of the aleatoric variability of the process (see Figure 1.4), the study produces a set of maps, presented here as upper and lower uncertainty bounds (typically 5th and 95th percentiles) of the vent opening probability at each location. These probability distribution maps are substantially robust with respect to different density estimation methods and expert aggregation models. Given the approach we have followed, our present results could be modified by eliciting the views of a group of experts composed of those who may hold different views from those who participated in this study, but we would be surprised if their findings diverged greatly from ours when the common basis is the same data, knowledge and process understanding. Of course, our own judgments could be modulated by any substantial new dataset, information or interpretation of the Campi Flegrei history and dynamics that might become available in the future.

Our results show evidence for a principal high probability region in the central-eastern portion of the caldera characterized by mean probability values of vent opening per km^2 that are about six times greater than the baseline value for the caldera. Significantly lower secondary maxima are found to exist in both the eastern and western parts of the caldera, with probabilities up to about 2-3 times larger than baseline. Nevertheless, the underlying spatial distribution of vent opening position probability is widely dispersed over the whole NYT caldera, including the offshore portion. Most importantly, we accompany our probabilities with quantified epistemic uncertainty estimates which are indicative, typically, of relative spreads ±30% of the local mean value, but with variations between approximately ±10% and ±50%, depending on the location. Notwithstanding the several assumptions and limitations of the analysis described above, the maps represent crucial input information for the development of quantitative hazard and risk maps of eruptive phenomena in the Campi Flegrei and can also be the basis for the generation of up-dated short-term vent opening probability maps, once monitoring information in an impending eruption becomes available.

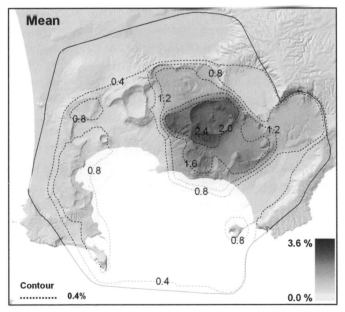

Figure 1.4. Mean probability map $E^E[\mu_{\tilde{X}}(\cdot)]$ of vent opening location conditional on the occurrence of an eruption, modified from [17]. Reported values indicate the percentage probability of vent opening per km^2.

1.3. Pyroclastic density current invasion maps

This chapter will focus on the definition of quantitative probabilistic PDC invasion hazard maps for the Campi Flegrei area conditional on the occurrence of an explosive eruption, encompassing the probabilistic assessment of potential vent opening locations derived in the previous chapter.

Pyroclastic density currents (PDCs) are probably the least predictable and the most dangerous of all volcanic hazards. As a consequence they have been responsible for most deaths in volcanic eruptions in recent times and they present the most important challenge of all volcanic hazards for disaster planners at volcanoes in densely inhabited regions. PDCs are laterally moving, buoyantly expanding mixtures of hot gases and fragmental particles (ash, lapilli, blocks, and boulders); indeed the word pyroclast is derived from the Greek $\pi\nu\rho$, meaning 'fire', and $\kappa\lambda\alpha\sigma\tau\sigma\varsigma$, meaning 'broken in pieces'.

Plinio il Giovane was the first to describe the phenomenon as 'flames of fire' that destroyed Pompeii and Herculaneum in the famous eruption of Mount Vesuvius in AD 79. The first direct scientific observations and descriptions of a pyroclastic flow, at that time called 'nuée ardente' (glowing cloud), occurred with the eruption of Montagne Pelée (Martinique) in 1902, when the town of St. Pierre was completely razed causing about 28,000 fatalities. An accelerated interest followed to the extensive observations of a large explosive eruption at Mount St Helens (USA) in 1980 (see Figure 1.5). This event led to a new understanding of highly mobile and intensively destructive PDC produced by a lateral volcanic blast, in addition to the documentation of the currents generated by column collapses. An even closer observation of dangerous pyroclastic flows occurred during the dome-forming and sporadically explosive eruptive activity of the Soufrière Hills volcano on the small island of Montserrat (UK) in the Caribbean sea (see Figure 1.6); it began in 1995 and has continued through 2010, producing currents of variable types and scales. Several villages including the island capital town Plymouth, were destroyed by these flows, thus allowing a first modern observation and measurement of the impacts produced by these currents on urbanized areas and people engulfed outdoors.

The eruptions of El Chichon (Mexico) in 1982, Unzen volcano (Japan) in 1991-1995 and Mt. Pinatubo in 1991, as well as the recent eruptions of Merapi (Indonesia) of 1990s, 2006 and 2010 offer other well documented examples of hazardous PDCs with remarkable impacts on nearby territory and populations. See [61, 116, 36] and [23] for more detailed hazard scenarios and additional information also concerning the following paragraphs.

Figure 1.5. Initial explosion produced by the opening blast of May 18, 1980 eruption of Mt. St. Helens; from [116].

PDC invasion hazard represents an extreme danger for the regions exposed to them: preventive evacuation of territory threatened by this hazard in generally required when the risk is high. In addition to the obvious threat represented by the capability of voluminous currents to bury the areas invaded under thick layers of pyroclastic deposits, the main hazards on land are represented by their dynamic pressure (average kinetic energy per unit volume), lethal temperature, fine ash-in-air concentrations, and the capacity to incorporate and transport loose rocks, trees and construction material that contribute added impact forces. These sources of hazard act in concert, *e.g.* with building damage exposing inhabitants to hot ash. The heat combined with the high density of material within pyroclastic flows obliterates objects in their paths, making them the most destructive of the volcanic hazards: the property damage is severe for even small dilute currents and there are very few survivors amongst those caught in their path. In addition there are no clear precursors to the production of a PDC, and multiple currents can be produced with flow directions that are completely random.

High uncertainty in the hazard leads to difficulties in the decision making process and forces emergency managers to make conservative judgements in terms of exclusion zones and evacuations. Populations must be evacuated from hazard zones prior to the onset of a PDC, infrastructure damage cannot be avoided, and currently no prevention measures are capable of protecting buildings and lifelines. In addition, the large amounts of hot ash deposited onto the landscape and particularly into channels extending from the volcanoes increases the risk of landslides and lahars: the ash-rich nature of such material means that it can be re-mobilized easily for decades after an eruption and then extend into jurisdictions that were unaffected by the actual eruption. Finally, additional complex hazards are associated with PDCs generated in the proximity of the sea or a lake (as may happen in Campi Flegrei). The denser component of the flow submerges and transforms to a long run-out submarine debris flow, while the more dilute part of the flow moves as a hot sandy cloud over water, sometimes for great distances. Moreover, the explosive interaction of seawater with blocky hot flows may cause an inland-directed base surge, and the impact of the PDC into the water body may generate a tsunami hazard affecting even distant islands and shorelines.

The generation of PDCs can be caused by several volcanic processes verified by different types of eruption, but the most common are:

- the gravitational collapse of a portion of the volcanic column due to reduction of efficiency of air entrainment into the hot, ash laden plume;
- the gravitational collapse and disintegration of a gas-enriched crystallized lava dome (*i.e.* a mound-shaped protrusions of viscous lava, see [26]) or flow front;
- the sudden explosion of a gas-enriched dome or cryptodome (lateral blast).

Depending on the scale of the explosive events, these mechanisms can produce PDCs with volumes ranging from a less than a cubic kilometer up to tens or hundreds. The larger ones can travel more than one hundred kilometers, although none on that scale have occurred for several thousand years. The density, velocity and temperature of the current are governed by the complex interactions between pyroclastic particles and the highly turbulent gaseous flow.

The pyroclastic particles are generated by the fragmentation of the bubbly magma or hot vesicular dome rock during its decompression, and by erosion; their distribution within the current can change drastically as a function of eruption mechanisms. During the gravity-influenced flow propagation, pyroclastic particles tend to segregate accordingly to their size, shape, and density: a steep density stratification is produced with solid concentrations ranging from dense packing at the base to very dilute near the top flow boundary. Particles are mostly suspended by turbulence in the more diluted part of the current, whereas in the basal layer they are mainly supported by fluid pressure and collisions. This view includes the two end-members of the traditional field-based classification of pyroclastic deposits: pyroclastic surges are dilute, thick and energetic currents

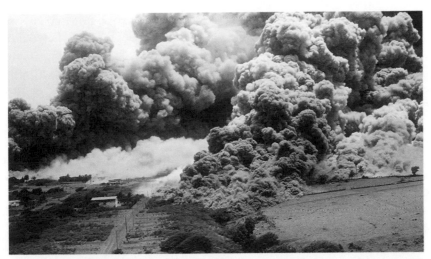

Figure 1.6. PDC during the final minutes of the small dome collapse of June 25, 1997 at Soufrière Hills Volcano, Montserrat; from [36].

that can decouple from a more concentrated underlying current and are able to traverse topographical obstacles at high speed and temperatures, possibly traveling beyond the limits of the basal pyroclastic flow; PDCs sensu stricto are the denser, thinner and concentrated currents that typically move below the dilute surge and are more strongly influenced by topography.

The PDC hazard assessment and zonation have been traditionally defined on the basis of the recognized extent of pyroclastic deposits of flows and surges of historical age, and the reconstruction of prehistoric deposits. The technique is limited in that deposit preservation is often poor, particularly for surges, and differences in vent location, edifice height and channel topography at the time of prior eruptions may be unknown, but could greatly affect the potential extent of a future current. Thus computer models based on current topography were first recognized as useful to check deposit based hazard boundaries, and with more sophisticated advancement became a primary approach for hazard mapping. In parallel, theoretical and experimental investigations have made significant progress with the development of a variety of physical and numerical models of PDCs. Simplified 1D/2D flow models and statistically based representations of the flows neglect several physical processes as well as 3D and transient effects of their dynamics. Nevertheless, once tuned with available field data sets, they enable helpful modelling during crises. They also could be adapted to Monte Carlo simulations to investigate model sensitivity to input parameters, such as flow volume and vent location, and to produce probabilistic hazard maps as we will show in the sequel. Alternatively, multidimensional and multi-phase flow models are becoming increasingly effective in representing the complex behaviour of such phenomena and enable simulations that include remarkable details of the scenario conditions. However currently they cannot yet be readily adapted to Monte Carlo simulations due to the excessive computing time required.

Despite significant progress, assessments of PDC hazard are still influenced by a remarkable amount of uncertainty: the quantification and communication of the diverse uncertainty sources appear to be the main challenge. The physical formulation of the process dynamics is currently not well constrained, furthermore initial and boundary conditions are subject to large epistemic and aleatoric variabilities that are just partially constrained by direct observations; the scale and precise source locations for potential threatening currents are generally poorly determined, and even the digital elevation map (DEM) representation of topography may only crudely approximate the topography at the time of a future hazardous event (*e.g.* [114, 66, 32, 33]). In Campi Flegrei caldera settings, the study of the PDC invasion hazard is additionally complicated by the remarkable variability of potential vent locations and eruption scales as well as by the complex dynamics of flows over a strongly heterogeneous topography (see [150]). Current efforts of the scientific community are aimed at estimating PDC hazards on a fully probabilistic basis, taking into account the aleatoric physical variabilities as well as the relevant epistemic uncertainties affecting the process: this would represent an important step toward the long-term goal of developing an interdisciplinary and integrated approach to risk reduction.

1.3.1. Formal definition of the PDC invasion maps

We characterized the hazard in terms of area invaded (or inundated) by the next PDC phenomenon. The definition of the probability measure representing such area and its explicit construction separating epistemic uncertainty from physical variability as we did for the map of vent opening, is one of the purposes of this chapter.

Definition 1.4 (The distribution of PDC invaded areas). Let Y be a random variable from the sample space $(\Omega, \mathcal{F}, P)$ to $(\mathbb{R}_+, \mathcal{B}(\mathbb{R}_+))$, representing the area invaded by PDCs during the next explosive eruption at Campi Flegrei. Let ν_Y be the probability measure that is the law of Y on $(\mathbb{R}_+, \mathcal{B}(\mathbb{R}_+))$; it is called a distribution of PDC invaded areas.

To construct a doubly stochastic model for this variable, we follow again the nontrivial probability structure of Definition 1.1, projecting the space $(W, \mathcal{W}, M)$ on $(\mathbb{R}_+, \mathcal{B}(\mathbb{R}_+))$.

Definition 1.5 (Conditional distribution of PDC invaded areas). Let π^2 be a measurable function from $(W, \mathcal{W}, M)$ to $(\mathbb{R}_+, \mathcal{B}(\mathbb{R}_+))$, representing the projection of the physical space onto the eruptive scale space. We assume that

$$Y(\omega) = \pi^2\left(\chi(\xi(\omega), \omega)\right), \text{ for almost every } \omega \in \Omega,$$

and we define the random variable $\check{Y}$ from $(E \times \Omega, \mathcal{E} \otimes \mathcal{F}, \eta \otimes P)$ to $(A, \mathcal{B}(A))$ as

$$\check{Y}(e, \omega) := \pi^2\left(\chi(e, \omega)\right).$$

For each $e \in E$ the random variable $\check{Y}(e, \cdot)$ on $(\Omega, \mathcal{F}, P)$ represents the area invaded by the next PDC phenomenon at Campi Flegrei once adopted the epistemic assumption e. Its law $\nu_{\check{Y}}(e)$ is called probability distribution of PDC invaded area conditional on the epistemic assumption e.

In particular $\nu_{\check{Y}} = \pi_\sharp^2(M)$: similarly to Definition 1.3 it is a well defined random measure on $(A, \mathcal{B}(A))$, parameterized on E. In principle we assume the independence of X and Y, and, for each $e \in E$, the independence of $\check{X}(e, \cdot)$ and $\check{Y}(e, \cdot)$: anyways a possible correlation between the location of the next vent and the area invaded by a PDC originating from it is debated. Of the main importance in this chapter is the definition of a simple model representing the propagation of a PDC of a particular scale, once assumed its eruptive vent location.

Definition 1.6 (The simplified flow model). Let $B \subseteq \mathbb{R}^2$ be a domain such that $B \supseteq A$, representing an enlarged zone possibly affected by PDC hazard, considered with its Borel sigma algebra $\mathcal{B}(B)$. Let F be a functional from $A \times \mathbb{R}_+$ to the Borel subsets of B, such that $F(x, y)$ represents the set invaded by a PDC propagating from a vent x with a scale y.

Combining the map of vent opening μ_X with the distribution of PDC invaded areas ν_Y and by using the simplified propagation model F, it is possible to produce probabilistic hazard maps of PDC invasion, estimating the effects of the considered sources of epistemic uncertainty.

Definition 1.7 (The maps of PDC invasion probability). Let p be a measurable function from $(B, \mathcal{B}(B))$ to $([0, 1], \mathcal{B}([0, 1]))$, that is defined as

$$p := E[1_{F(X,Y)}]$$

and represents the probability of each point of B to be reached by the next PDC. For each $z \in B$ we also define a random variable $\check{p}$ from $(E, \mathcal{E}, \eta)$ to $([0, 1], \mathcal{B}([0, 1]))$ as

$$[\check{p}(z)](e) := E[1_{F(\check{X}(e, \cdot), \check{Y}(e, \cdot))}](z).$$

It expresses the probability of each point of B to be reached by the next PDC, conditional on the epistemic assumption e.

With $\check{p}$ we estimate the random probability of each point of B to be reached by the next PDC as a function of $e \in E$. This is calculated by a double Monte Carlo simulation with a nested structure.

1.3.2. Description of the achievements

Through the application of the doubly stochastic model we produce the first quantitative background (or long-term/base-rate, *i.e.* in conditions of no unrest) probabilistic maps of PDC invasion hazard able to incorporate some of the main sources of epistemic uncertainty that influence the models for aleatoric (physical) variability (see Figure 1.7). In particular, by a Monte Carlo simulation approach the new method developed combines the spatial probability distribution of vent opening locations, inferences about the spatial density distribution of PDC invasion areas informed by reconstruction of deposits from eruptions in the last 15 ka, and a simplified PDC model able to describe the pyroclastic flow kinematics and to account for the main effects of topography on flow propagation. In Chapter 5 we include a digression about the physical details of such model and its implementation (see also [85, 86, 80, 81]). In addition our mapping attempts to quantify, relying again on the formal structured expert judgement approach, some other relevant sources of epistemic uncertainty in addition to the location of future vent opening: like the reconstruction of the dispersal of PDC deposits, or the possibility that a future eruption could be characterized by the opening of two simultaneous vents located perhaps several kilometers apart, as highlighted by [97] for the Averno 2 and Solfatara eruptions.

Our results clearly suggest that the entire caldera has potential to be affected, with a mean probability of flow invasion higher than about 5% and the central-eastern area of the caldera (*i.e.* Agnano-Astroni-Solfatara) having invasion probabilities above 30% (with local peaks at or above 50% in Agnano). Significant mean probabilities (up to values of about 10%) are also computed in some areas outside the caldera border (*i.e.* over Collina di Posillipo and in some neighborhoods of Naples). Our findings are robust against different assumptions about several of the main physical and numerical parameters adopted in the study. In addition to mean values of probability of PDC invasion, this study provides the first estimates of the credible uncertainty ranges associated with such probability estimates in relation to some key sources of epistemic uncertainty. From our analysis, the typical uncertainty ranges affecting invasion probabilities inside the caldera lay between ±15 and ±35% of the

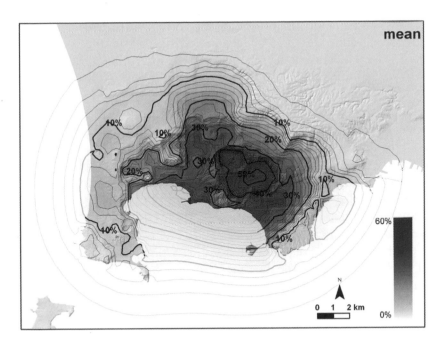

Figure 1.7. Mean probability map $p(z) = E^E[[\breve{p}(z)](\cdot)]$ of PDC invasion hazard conditional on the occurrence of an explosive eruption, modified from [118]. The map assumes that PDCs originate from a single vent per eruption, and that the vent is located in the on-land part of the caldera.

local mean value, with an average of about ±25%; wider uncertainties are found outside the caldera, with an average above ±50% and a significantly larger variability from place to place. Despite the several assumptions and limitations of this study, including the partial subjectivity of the approach followed, such first estimates of epistemic uncertainty provide crucial information that needs to be carefully accounted for quantifying the likelihood of PDC hazards, and risks, associated with a future eruption occurring at Campi Flegrei. A scientific report about this study have been presented to Dipartimento della Protezione Civile and Commissione Nazionale Grandi Rischi (see [117]) to provide additional information for the re-definition of the Red Zone at Campi Flegrei, the area considered highly at risk from pyroclastic flows and which would need to be evacuated before an eruption.

1.4. Time-space model for the next eruption

Temporal scale estimation has been deliberately left out from the previous analyses because of its complexity: the maps provided represent exclusively spatial distributions conditional on the occurrence of a new (explosive) vent opening. This chapter will cope with the construction of a robust temporal model capable of producing a background (long-term) probability distribution for the time of the next explosive eruption at Campi Flegrei.

The known sequence of eruptive events is remarkably non-homogeneous, both in time and space (*e.g.* [133, 96]). Indeed activity has been subdivided into three distinct epochs (*i.e.* Epoch I, 15 - 10.6 ka, Epoch II, 9.6 - 9.1 ka, and Epoch III, 5.5 - 3.8 ka BP, estimates from [145]), alternated by long periods of quiescence (see Figure 1.8), and the stratigraphic record shows the presence of clusters of eruptions in time-space inside the single epochs of activity. The record of past eruption times is affected by a large epistemic uncertainty: only for a few of them datation ranges have been estimated, while for the most only the stratigraphic order has been assessed; even the times and durations of the eruptive epochs and of the periods of quiescence are very uncertain. It is fundamental that

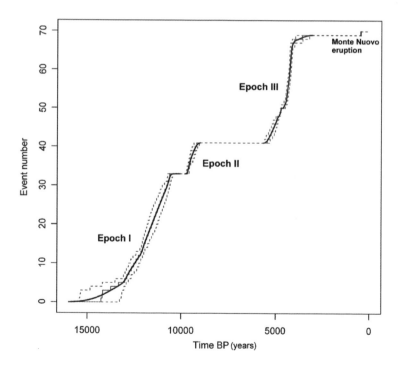

Figure 1.8. Cumulative event number as a function of time during the entire eruptive record of the last 15 ka. The bold line is the mean value, the narrow line is the 50th percentile and the dashed lines are 5th and 95th percentiles of the epistemic uncertainty.

the opening of a vent in a particular location and at a specific time seems to increase the probability of another vent opening in the nearby area and in the next decades-centuries (self-exciting effect): for this reason the time-space mathematical model that has been developed takes into account both the quantification of the significant uncertainty affecting the eruptive record and the possible self-exciting behaviour of the system. This kind of study is crucial also to understand the likelihood and the consequences of Monte Nuovo in AD 1538. The meaning of this event is indeed controversial because it is separated by 2.7-3.7 ka from the last event of epoch III and followed by 477 years without explosive activity. In particular it could represent, adopting the past behaviour of the volcano as valid, the first eruption of an incoming new eruptive epoch.

1.4.1. Formal definition of the time-space model

The available temporal information about past eruptive events consists of their ordered stratigraphic sequence (unsure or even unknown in a few cases) and some large datation windows for a subfamily of them. We follow again Definition 1.1 for assessing the epistemic uncertainty.

Definition 1.8 (Time-space record with uncertainty). Let $(w_i)_{i=1,...,n}$ be the set of all the eruptive events considered. Assume that τ is a random variable from $(E, \mathcal{E}, \eta)$ to the space $\mathcal{S}(n)$ of the permutations of $\{1, \ldots, n\}$ such that $(v_j)_{j=1,...,n}$, where

$$v_j := w_{\tau(j)}, \quad \forall j,$$

represents a random sample for the ordered family of eruptive events. Let $(t_j)_{j=1,...,n}$ be a vector of real random variables from $(E, \mathcal{E}, \eta)$ to $\mathbb{R}^n_+$, each t_j representing the time of eruptive event v_j, consistent with the datation bounds available. For each $j = 1, \ldots, n$ let V_j be a random variable from $(E, \mathcal{E}, \eta)$ to $(A, \mathcal{B}(A))$ representing the location of the eruption v_j. We define the random set of random variables

$$\Theta_l := \{t_j \ : \ V_j \in A_l\}$$

representing the times of each eruption v_j that occurred in the zone A_l. We adopt the notation $\Theta_l = (t_j^l)_{j=1,...,n_l}$.

Based on this, we define a family of counting processes representing the number of vents opening in each zone of the caldera as a function of time. The model adopted relies on a 'Cox-Hawkes process', *i.e.* a doubly stochastic Hawkes process (see [44, 46, 49]), including a spatial localization in the different sectors of the caldera. The Hawkes processes are non-homogeneous Poisson processes (NHPP) in which the intensity rate increases with a jump whenever an event occurs and instead decreases (often following exponential or sigmoid decay curves) as time passes without any event occurring; see [15] for an example of Hawkes process adoption in volcanic hazard assessment. The Cox processes are simply the doubly stochastic version of the NHPP, in which the model parameters are assumed affected by uncertainty; see [88, 89, 90] for some applications of Cox processes in volcanology. The innovative model developed presents both these properties; in particular we explored the case of an exclusively local self-excitement, *i.e.* without interaction between different zones.

Definition 1.9 (The Cox-Hawkes process). Let $Z = (Z^l)_{l=1,\dots,N}$ be a doubly stochastic multivariate Hawkes process on $(\Omega, \mathcal{F}, P)$, adopting the nontrivial structure of Definition 1.1. Let φ be an application from E to the functional space of continuous decreasing functions on $\mathbb{R}_+$, representing the diminishing of self interaction for the process Z. For each $l = 1, \dots, N$, let λ_0^l be a random variable on $(E, \mathcal{E})$ representing the base rate of the process Z^l. The intensity function of the component Z_l is then expressed by

$$\lambda^l(t, \omega) = \lambda_0^l(e) + \sum_{t_i^l(w)<t} [\varphi(e)](t - t_i^l(\omega))$$

$$= \lambda_0^l(e) + \int_0^t [\varphi(e)](t - u) dZ_u^l(\omega), \quad \forall l = 1, \dots, N,$$

where we assume $e = \xi(\omega)$ of Definition 1.1.

The main problem is to cope with the assessment of the function φ, also depending of epistemic uncertainty: indeed we developed a mathematical procedure based on maximizing the likelihoods of random sampled past records inside a Monte Carlo simulation. This is aimed at finding an uncertainty distribution for the physical parameters of the model: the base rate λ_0, the time scale of excitement decay T and the mean number of offspring events μ; each of them represented as a random variable on the space $(E, \mathcal{E}, \eta)$.

Definition 1.10 (Conditional Cox-Hawkes processes). Let π^3 be a measurable function from $(W, \mathcal{W}, M)$ to the space of l-dimensional counting measures, representing the projection of the physical space onto the set of next eruptions times in each of the caldera zones. We assume that

$$Z(\omega) = \pi^3 \left(\chi(\xi(\omega), \omega) \right), \text{ for almost every } \omega \in \Omega,$$

and we define the point process $\check{Z}$ from $(E \times \Omega, \mathcal{E} \otimes \mathcal{F}, \eta \otimes P)$ to $(A, \mathcal{B}(A))$ as

$$\check{Z}(e, \omega) := \pi^3 \left(\chi(e, \omega) \right).$$

For each $e \in E$ the point process $\check{Z}(e, \cdot)$ on $(\Omega, \mathcal{F}, P)$ represents the set of next eruptions times at Campi Flegrei once adopted the epistemic assumption e.

But the main purpose of this chapter is the time scale assessment for the next future eruption at Campi Flegrei. In the following definition, with the minimum of a real point process we will indicate its minimum jump time.

Definition 1.11 (The next eruption time distribution). Let Z_{mn} be a multivariate Cox-Hawkes process representing eruptions in each of the caldera zones, and starting from a situation without excitement except for the residual additional intensity from an event occurred $t_0 = 477$ years before time 0, in zone 3 (Averno-Monte Nuovo). Then define on $(\Omega, \mathcal{F}, P)$ the real positive random variable

$$Z^* := \min_l Z^l_{mn},$$

representing the remaining time before the next eruption at Campi Flegrei. Let ϱ_{Z^*} be the probability measure that is the law of Z^* on $(\mathbb{R}_+, \mathcal{B}(\mathbb{R}_+))$; it is called a distribution of next eruption time.

In particular it is possible to use the doubly stochastic structure of Z_{mn} and define a conditional version of this variable.

Definition 1.12 (Conditional next eruption time distribution). For each $e \in E$ let $\check{Z}_{mn}(e, \cdot)$ be a conditional multivariate Cox-Hawkes process representing eruptions in each of the caldera zones, and starting from a situation without excitement except that from an event occurred $t_0 = 477$ years before time 0, in zone 3. Then define on $(E \times \Omega, \mathcal{E} \otimes \mathcal{F}, \eta \otimes P)$ the real positive random variable

$$\check{Z}^*(e, \omega) := \min_l \check{Z}^l_{mn}(e, \omega),$$

and let $\varrho_{\check{Z}^*}$ be its law on $\mathbb{R}_+$. For each $e \in E$ the random variable $\check{Z}^*(e, \cdot)$ on $(\Omega, \mathcal{F}, P)$ represents the remaining time before the next eruption at Campi Flegrei once adopted the epistemic assumption e. Its law $\varrho_{\check{Z}^*}(e)$ is called probability distribution of next eruption time conditional on the epistemic assumption e.

The random variable $\check{Z}^*$ is assessed through a double Monte Carlo simulation with a nested structure, similarly to the previous cases. It will be represented trough the values of its density function on $\mathbb{R}_+$.

1.4.2. Description of the achievements

Relying on the past behaviour of the volcano, the doubly stochastic time-space model adopted allows to simulate sequences of future eruptions as well as to better understand the spatial and temporal behaviour of the system. Results confirm qualitative appreciations of time-space clustering in the Campi Flegrei with some differences between the different epochs of activity, in particular the Epoch I record produces a mean rate of generation for the new clusters of one on 148 years; Epoch II record produces a mean rate of one on 63 years, Epoch III of one on 106 years; these values are affected by a relevant epistemic uncertainty. The duration of the self-excitement from an event appears quite different between the epochs, with an average estimate of 658 years for Epoch I, 101 years for Epoch II and 96 years for Epoch III.

It is confirmed that the eruption of Monte Nuovo is an anomaly compared to the prevalent pattern of previous epochs, if we do not pose other assumptions. In particular assuming Monte Nuovo as the opening event of a new epoch, the likelihood of having observed 477 years without explosive activity is below 5% in average, but it increases above 30% if we consider separately the record of the western part of the caldera. We remark that the probability for each new eruption of developing a cluster of subsequent events is around 35% in average if we consider the whole record, but drops around 15% for the western sub-record: the clustering behavior is much weaker in that case.

There are several volcanological assumptions that could be made: each of them corresponds to a different forecast obtained with Monte Carlo simulations for the timing of the next eruption at Campi Flegrei. In Figure 1.9 are shown the mean probability distributions associated with different volcanological assumptions. Considering the three eruptive epochs as independent samples and assuming Monte Nuovo as the first event of a new epoch of activity, we obtain a mean estimate of 103 years from the present time (year 2015), with a physical variability 5th and 95th percentiles of 5 and 318 years respectively. Each of these values is affected by epistemic uncertainty, which have been estimated as $\pm 30\%$ and slightly skewed towards the positive side. Optionally considering only the starting phases of the epochs before climactic events as Agnano Monte Spina or Pomici Principali slightly increases these estimates.

Relying on separate records for eastern and western sectors activity of the Campi Flegrei caldera, we obtain that the eastern sector dataset is still quite consistent with the previous estimates, but the western instead rises them of almost five times to 470 years in mean from the present time, with physical variability percentiles of 25 to 1467 years, with an epistemic uncertainty estimated as ±35% again skewed towards the positive side. Also assuming a unique sequence of events including also the periods of quiescence between the eruptive epochs produces much larger time estimates than considering only the eruptive epochs; however the probability model in this case seems too simple for re-producing both the long periods of quiescence and the more frequent activity inside eruptive epochs.

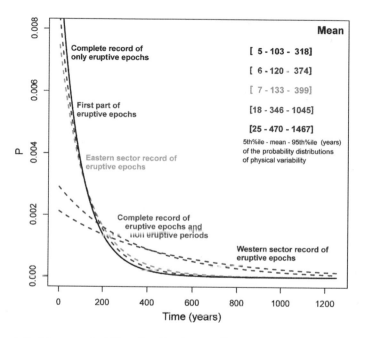

Figure 1.9. Mean probability distributions $E^E[\varrho_{\dot{Z}*}(\cdot)]$ for the remaining time before the next eruption. The curves indicate the probability density function per year. The values reported are the mean with respect to epistemic uncertainty of the 5th and 95th percentiles and the mean value of the physical variability. Different colours correspond to alternative geological assumptions.

Further geological studies aimed at the reduction of the number of events without datation bounds during Epoch I and Epoch II could improve the cluster recognition, with the possibility of obtaining an eruption pattern similar to one of Epoch III also during the previous epochs. Moreover, additional research for better understanding the mechanisms of reactivation of calderas after long periods of quiescence confirms to be of the main importance, together with the exploration of the dissimilarities between the western and eastern sectors of the Campi Flegrei caldera, also confirmed by the remarkably different erupted volumes and geological structures involved.

Notation 1.13. In the following chapters we will recall the most of the definitions of this introduction, with a new reference number suitable with their sequence.

Chapter 2
Vent opening probability maps

2.1. Summary

Campi Flegrei is an active volcanic area situated in the Campanian Plain (Italy) and dominated by a resurgent caldera. The great majority of past eruptions have been explosive, variable in magnitude, intensity and in their vent locations. In this chapter we present a probabilistic analysis using a variety of volcanological datasets to map the background spatial probability of vent opening conditional on the occurrence of an event in the foreseeable future. The analysis focuses on the reconstruction of the location of past eruptive vents in the last 15 ka, including the distribution of faults and surface fractures as being representative of areas of crustal weakness (see Figure 2.1). One of the key objectives is to incorporate some of the main sources of epistemic uncertainty about the volcanic system through a structured expert elicitation, thereby quantifying uncertainties for certain important model parameters and allowing outcomes from different expert weighting models to be evaluated.

Results indicate that past vent locations are the most informative factors governing the probabilities of vent opening, followed by the locations of faults and then fractures. The vent opening probability maps highlight the presence of a sizeable region in the central-eastern part of the caldera where the likelihood of new vent opening per km^2 is about six times higher than the baseline value for the whole caldera. While these probability values have substantial uncertainties associated with them, findings provide a rational basis for hazard mapping of the next eruption at Campi Flegrei caldera. The definition of a vent opening probability map and its explicit construction, estimating epistemic uncertainties and their influence on the physical variability, is the purpose of this chapter. We based our probability model on the abstract definition of the vent opening location as a random variable X, associated with a probability measure μ_X on the domain of Campi Flegrei.

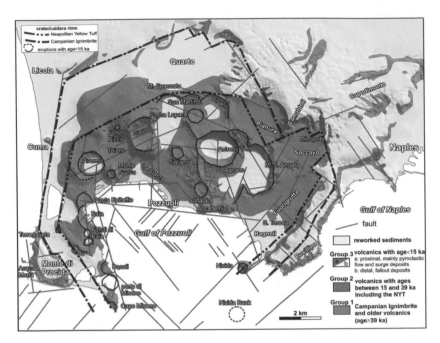

Figure 2.1. Simplified geological map of Campi Flegrei caldera showing regional fault traces and main morphological structures such as caldera and crater rims and faults derived from sea seismic profiles; from [17].

Definition 2.1 (The vent opening probability map). Let $A \subseteq \mathbb{R}^2$ be a domain representing the area of the volcanic system, considered with its Borel sigma algebra $\mathcal{B}(A)$. Let X be a random variable from the sample space $(\Omega, \mathcal{F}, P)$ to $(A, \mathcal{B}(A))$, representing the location of the next eruptive vent at Campi Flegrei. Let μ_X be the probability measure that is the law of X on $(A, \mathcal{B}(A))$; it is called a map of vent opening.

For assessing an estimation of the unknown measure μ_X we quantified its epistemic uncertainty defining a random measure. We adopted the doubly stochastic probability structure and the notation of Definition 1.1, just projecting the physical space $(W, \mathcal{W}, M)$ on the spatial domain $(A, \mathcal{B}(A))$.

Definition 2.2 (Conditional vent opening probability map). Let π^1 be a measurable function from $(W, \mathcal{W}, M)$ to $(A, \mathcal{B}(A))$, representing the projection of the physical space onto the space of the vent opening location. We assume that

$$X(\omega) = \pi^1\left(\chi\left(\xi(\omega), \omega\right)\right), \text{ for almost every } \omega \in \Omega,$$

and we define the random variable $\check{X}$ from $(E \times \Omega, \mathcal{E} \otimes \mathcal{F}, \eta \otimes P)$ to $(A, \mathcal{B}(A))$ as

$$\check{X}(e, \omega) := \pi^1 (\chi(e, \omega)).$$

For each $e \in E$ the random variable $\check{X}(e, \cdot)$ on $(\Omega, \mathcal{F}, P)$ represents the location of the next eruptive vent at Campi Flegrei once adopted the epistemic assumption e. Its law $\mu_{\check{X}}(e)$ is called vent opening probability map conditional on the epistemic assumption e.

Remark 2.3. Defining likely locations of future vents is a key scientific goal for hazard and risk assessment, especially given the wide dispersion of past eruptive vents within the caldera. Alberico *et al.* [3] presented a first quantitative analysis, based on seven geophysical, geological and geochemical parameters, each one assumed to be representative of a degree of anomaly. These parameters were combined to produce a spatial distribution of the probability of vent opening on a regular grid with cells of side 1 km, covering the whole caldera. Their findings suggested that the inner portion of the caldera (approximately a circular area with a diameter of about 6 km centered on the town of Pozzuoli) had the highest probability of vent opening. In contrast, Orsi *et al.* [122] assumed, mostly on structural considerations, that the chances of a new vent opening depended only on the distribution of past vents of Epoch III. They qualitatively identified two distinct areas, one with higher probability of vent opening (approximately located in the region of Astroni, Agnano and part of San Vito) and the other with lower probability (approximately located in the area of Averno and Monte Nuovo). More recently, Selva *et al.* [141] produced a probabilistic map, over a regular grid with cells of sides 500 m, based on a Bayesian inference procedure and reporting uncertainty ranges for probability values (see also Appendix B of this chapter). Their approach included information on the location of past vents of Epoch III, starting from a prior distribution defined by assigning scores to the presence of tectonic structures or eruptive vents of the last 15 ka in the NYT caldera. This study highlighted how the probability of vent opening is widely distributed over the caldera, with two areas of higher probability of vent opening located in the Agnano-Astroni-San Vito and the Averno-Monte Nuovo areas.

2.2. Methodology

We followed a structured expert elicitation and judgment pooling approach (*e.g.* [41, 7], and Chapter 5) to quantify epistemic uncertainties on evidence coming from different strands of volcanological data and then merge these distributions to produce a doubly stochastic probabilistic vent opening map that accommodates and expresses these different sources of uncertainty. Our method is based on the assumption that the probability of new vent opening can be computed as a weighted linear combination of the spatial distributions of key physical variables of the system that reflect, or can influence, this volcanic process. Similar approaches, but involving different techniques, have been applied in [141] and [12] for mapping vent opening at explosive volcanoes and in [104, 27] and [37] for generating vent opening maps (also called susceptibility maps, *e.g.* [104]) at effusive volcanoes. A similar approach has also been applied for the generation of ensemble maps of seismic and tectonic hazards for planning geological areas suited for radioactive waste storage or disposal (*e.g.* [31]). According to the notation of Definition 2.2, we focused on the assessment of the random probability measure $\mu_{\check{X}}$ representing vent opening spatial distribution as a function of epistemic uncertainty.

Definition 2.4 (Random mixture of simple maps). Let $(X_i)_{i=1,\dots,d}$ be a family of random variables from $(\Omega, \mathcal{F}, P)$ to $(A, \mathcal{B}(A))$, and let $(\mu_i)_{i=1,\dots,d}$ be the family of their laws. They represent the spatial distributions of a number of different geological features supposed to be correlated to the opening of a new eruptive vent. We assume that μ_X can be expressed as the convex combination (with random coefficients) of such family. These d coefficients $\alpha = (\alpha_i)_{i=1,\dots,d}$ are random variables from $(E, \mathcal{E}, \eta)$ to $([0, 1], \mathcal{B}(0, 1))$ and the problem of assessing $\check{X}$ is reduced to find the distribution of their block:

$$\mu_{\check{X}}(e) = \sum_{i=1}^{d} \alpha_i(e)\mu_i, \quad \sum_{i}^{d} \alpha_i = 1.$$

We used data from literature and new data reported here. The variables considered in the analysis were: the distribution of the eruptive vents opened during the three epochs in the last 15 ka of activity of the volcano; the distribution of maximum fault dislocations, and the density of surface fractures over the whole caldera. Based on the present understanding of caldera systems, these five distributions, representative of the aleatoric variability of the vent opening process, appear to be the ones most closely correlated with vent opening potential, with faults and fractures representative of near-surface regions of crustal weakness in the caldera. We acknowledge that the probability of new intra-caldera vent opening could be correlated with other system variables or processes that we did not consider due to lack of knowledge about them. To account for any contribution from these neglected factors and to represent missing information, we included a conservative spatial uniform distribution inside the NYT caldera. The analysis focused on events from the last

15 ka of activity of the volcano since these are by far the best known and, given the volcanological and structural evolution of the caldera (see [132, 57, 122]), are also the most relevant for this study. In fact we presumed that the caldera did not evolved significantly over this interval; moreover, some differential weighting, from the elicitation findings, was applied which tested the effect of placing more emphasis on the most recent data.

A key aspect of the study was the identification, and where possible the quantification, of some of the main sources of epistemic uncertainty that are associated with the available data and therefore need to be reflected in the final maps. In particular, in reconstruction from deposits the attendant uncertainty on location of related eruptive vents was considered, as were the number of past events which do not correspond to presently identified vents but which do exist in the stratigraphic evidences (so-called 'lost vents') and the uncertainty of linear weights $(\alpha_i)_{i=1,...,d}$ to be associated with the variables that contribute to the definition of the mapping. The construction of the random vector α is a too difficult problem to be coped with a direct expert judgement approach, and simple Monte Carlo simulation was implemented to calculate the distribution of α from the distribution of the elicitation responses to simpler questions β, and then to obtain a sample of $\mu_{\tilde{X}}$.

Definition 2.5 (The uncertainty on the linear weights).
Let $\beta = (\beta_j)_{j=1,...,d'}$ be a family of random variables defined on $(E, \mathcal{E}, \eta)$ and representing the uncertainty profiles associated to each answer value of the elicitation's target questionnaire. Let f be a measurable function from $(\mathbb{R}^{d'}, \mathcal{B}(\mathbb{R}^{d'}))$ to $(\mathbb{R}^d, \mathcal{B}(\mathbb{R}^d))$ such that

$$f(\beta(e)) = \alpha(e)$$

for each $e \in E$.

In particular with regards to the uncertainty of the linear weights and the unknown values of some other variables, we adopted a hierarchical logic tree of questions and different scoring rule models for pooling group judgments, including performance-based (see Appendix A of this chapter; [41, 7, 68, 31]) and equal-weight models. The procedure differs from previous studies where the weights were directly and deterministically assigned by the authors to variables with unknown values (*e.g.* [141, 12]).

2.3. The volcanological datasets

The construction of the probability measures $(\mu_i)_{i=1,...,d}$ of Definition 2.4 from the incomplete geological information available has a main relevance: in the following we include a discussion about the details of the procedures followed, based on simple empirical rules shared with experts of the field. Such inputs to the probabilistic maps consist of three different types of datasets: (1) the spatial distribution of vent opening locations in the three epochs of the last 15 ka; (2) the spatial distribution of maximum fault displacement and (3) the surface fracture density. Unless reported otherwise, all three variables were mapped on a regular grid of 100x100 cells of side 250 m, covering the whole caldera, with the lower left corner of the grid at $(415000, 4510000)$ WGS84 UTM Zone 33 coordinates. As the outer boundary of the analysis we considered the rim of the CI caldera as reconstructed in [153] since all vents of the last 15 ka were inside it and even faults and fractures outside this area appear old and not correlated with the most recent volcanic activity.

2.3.1. Distributions of past vents

The location of past vents represents the principal information to consider when constructing a vent opening probability map. Therefore this variable was investigated in depth trying to quantify the different sources of uncertainty that affect it. In particular we focused on the uncertain location of the vents which, in most cases, cannot be represented as precise points, and on the uncertain number of vents that might have existed but now are not visible (lost vents). The locations of vents for the eruptive events that occurred in Epochs I, II and III (Figures 2.2, 2.3) are indicated on the maps by circles or ellipses representing the area where the eruptive vent (or fissure) was probably located during the eruption. Inside these areas it was decided to avoid more detailed assumptions as distributing more weight near the center of the ellipses or on their boundaries (as it could seem reasonable in case of caldera collapse), and the uncertainty was in principle distributed uniformly. Each eruption is geologically associated with an elliptic subset of the spatial domain A where it is likely that the volcanic vent (or fissure) was located during it.

Definition 2.6 (The past vents locations). Let $\mathcal{V}$ be a discrete set which elements $(w_i)_{i=1,\ldots,n}$ represent volcanic eruptions at Campi Flegrei and for each $i = 1, \ldots, n$ let $D_i \subseteq A$ be a set representing the enlarged location of the eruption w_i. For each D_i let ζ_i be a uniform probability measure supported on that set.

In general small circles/ellipses indicate a good knowledge of the vent location, mostly based on the existence of a crater, the presence of other surface morphological features, or a well-exposed areal deposit distribution. Large circles/ellipses indicate large uncertainty in vent location due to burial or destruction by subsequent eruptions, or by the action of seawater inundating the caldera. Migration of a vent during the same erup-

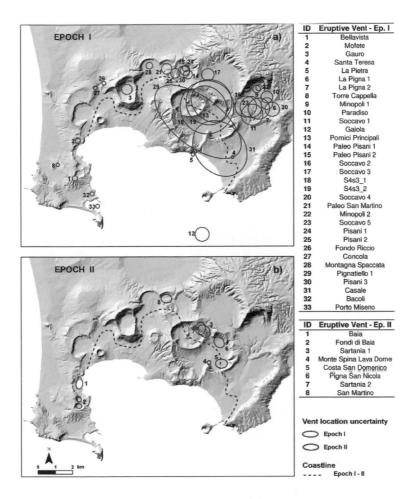

ID	Eruptive Vent - Ep. I
1	Bellavista
2	Mofete
3	Gauro
4	Santa Teresa
5	La Pietra
6	La Pigna 1
7	La Pigna 2
8	Torre Cappella
9	Minopoli 1
10	Paradiso
11	Soccavo 1
12	Gaiola
13	Pomici Principali
14	Paleo Pisani 1
15	Paleo Pisani 2
16	Soccavo 2
17	Soccavo 3
18	S4s3_1
19	S4s3_2
20	Soccavo 4
21	Paleo San Martino
22	Minopoli 2
23	Soccavo 5
24	Pisani 1
25	Pisani 2
26	Fondo Riccio
27	Concola
28	Montagna Spaccata
29	Pignatiello 1
30	Pisani 3
31	Casale
32	Bacoli
33	Porto Miseno

ID	Eruptive Vent - Ep. II
1	Baia
2	Fondi di Baia
3	Sartania 1
4	Monte Spina Lava Dome
5	Costa San Domenico
6	Pigna San Nicola
7	Sartania 2
8	San Martino

Vent location uncertainty
- Epoch I
- Epoch II

Coastline
- - - - Epoch I - II

Figure 2.2. Reconstruction of the location of the eruptive vents and fissures for the events occurred in (a) Epoch I and (b) Epoch II (from [17]). Numbered circles and ellipses indicate the assumed vent location of the events listed on the right side of the maps. The name of the events follows [145]. The dashed line indicates the likely location of the coast line between Epochs II and III (from [122]).

tive event was also considered and, where this was considered plausible, contributed to a large vent-location ellipse. We defined the vent location dataset by assuming a one-to-one relationship between the eruptive event (assumed as deposit erupted in a period of time representative of the eruption duration, *i.e.* of the order of days/months) and the eruptive vent from which it originated. The possible occurrence of eruptions with two simultaneously active vents in different sectors of the caldera was considered as two distinct events for the aim of vent zonation (see Chapter 3, for further considerations about this possibility). During Epoch I the recognized vents were mostly concentrated along the northern and eastern border portions of the caldera (Figure 2.2) whereas during Epochs II and III volcanism was mostly concentrated in the central-eastern part of the caldera (*i.e.* Agnano-Astroni-Solfatara; Figures 2.2, 2.3; [132, 57, 122, 97, 145]).

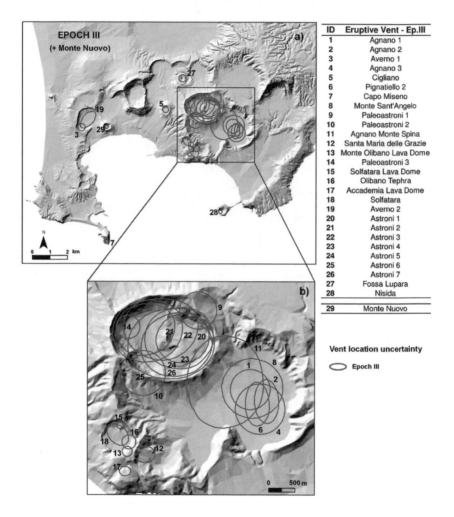

ID	Eruptive Vent - Ep.III
1	Agnano 1
2	Agnano 2
3	Averno 1
4	Agnano 3
5	Cigliano
6	Pignatiello 2
7	Capo Miseno
8	Monte Sant'Angelo
9	Paleoastroni 1
10	Paleoastroni 2
11	Agnano Monte Spina
12	Santa Maria delle Grazie
13	Monte Olibano Lava Dome
14	Paleoastroni 3
15	Solfatara Lava Dome
16	Olibano Tephra
17	Accademia Lava Dome
18	Solfatara
19	Averno 2
20	Astroni 1
21	Astroni 2
22	Astroni 3
23	Astroni 4
24	Astroni 5
25	Astroni 6
26	Astroni 7
27	Fossa Lupara
28	Nisida
29	Monte Nuovo

Figure 2.3. (a) Reconstruction of the location of the eruptive vents and fissures for the events occurred in Epoch III and of the Monte Nuovo eruption (from [17]). (b) The map represents an enlargement of the area of Agnano-Astroni-Solfatara where many events occurred. Numbered circles and ellipses indicate the assumed vent location of the events listed on the right side of the maps. The name of the events follows [145].

The three datasets of vent locations with respect to the three epochs of activity of the volcano (Figures 2.2 and 2.3) were the starting point for producing a first spatial distribution of probability of new vent opening, conditional on this information. We adopted two different approaches: a kernel density estimation with Gaussian distributions (Figure 2.4a), and a simpler probability distribution based on a partition of the caldera into finite zones (Figure 2.4b). It will be seen that the two approaches are complementary and produce quite consistent results. Both Figures 2.4a and 2.4b refer to the whole dataset of all vent locations from the three epochs, without discriminating between them. By contrast, in the generation of the final vent opening probability maps vents of different epochs are weighted differently, based on the outcomes of the elicitation.

The kernel density estimation is a non-parametric method for estimating the spatial density of future volcanic events based on the locations of past vents (*e.g.* [37, 15, 110]). Two important parts of the spatial density estimate are the kernel function and its bandwidth, or smoothing parameter. The kernel function can be any positive function K that integrates to one. In general, given a finite sample $X_i, i = 1, \ldots, N$, a kernel density estimator can be defined as:

$$f_h(x) = \frac{1}{N} \sum_{i=1}^{N} K \left(\frac{x - X_i}{h} \right)$$

where h is the bandwidth. K is assumed equal to a two-dimensional radially symmetric Gaussian kernel, as with many kernel estimators used in geologic hazard assessments (*e.g.* [38, 27, 110]). The bandwidth is typically selected using different theoretical and empirical methods developed for optimizing consistency with data (*e.g.* [62, 110]). Here we took it independently of the spatial location and equal to the mean minimum distance between the centers of the circles/ellipses for each separate epoch since the bandwidth is, in principle and other things being equal, related to the spatial spread of the observed past vents. A complication in our study is that the sample of past vent locations does not comprise points, but areas of uncertainty, and each vent area covers several cells of our grid, some of them completely, others only partially. Therefore for each cell we took into account the fraction of each vent area that it contains and then we applied the kernel convolution to this value. In addition, we also assumed that this kernel convolution does not spread the probability outside the CI caldera boundary. An advantage of this approach is that the spatial density estimate will be consistent with the spatial distribution of past volcanic events. A disadvantage of a symmetrical kernel function is that it does not explicitly allow for geological and structural boundaries and other directional volcanological information (see [37]). The areas with the highest density of past vents are those of Astroni and Agnano (maximum probability per km^2 respectively around 4.8% and 4.0%) followed by Soccavo, Solfatara and Pisani (Figure 2.4a).

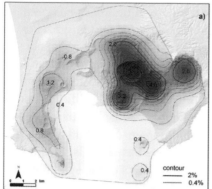 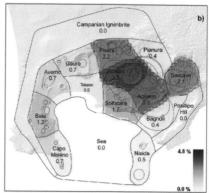

Figure 2.4. (a) Density distribution of the probability of vent opening obtained by using the vent location data of the three epochs of activity reported in Figures 2.2 and 2.3 and a kernel density estimation. Contour and colour values indicate the percentage probability of vent opening per km^2 (conditional on the occurrence of an eruption). (b) Density distribution of the probability of vent opening obtained by using the vent location data of the three epochs of activity reported in Figures 1.3 and 1.4 and the partitioning of the caldera in 16 homogeneous zones. Values reported in the different subareas indicate the percentage probability of vent opening per km^2 (conditional on the occurrence of an eruption). From [17].

The caldera partition approach was developed to take into account this last challenge and to complement the kernel based approach described above. We subdivided the whole Campi Flegrei caldera into 16 zones, each characterized by internally consistent geological and volcanological features and therefore a quasi-homogeneous distribution of vent opening frequencies. Considering also the geological information about caldera boundaries and time-space clustering of past eruptive events (see also Chapter 4), we produced a partitioning of the caldera. The following is a formal definition for it, for the sake of notation.

Definition 2.7 (The caldera partition). Let $(A_l)_{l=1,...,N}$ be a finite partition of the spatial domain A of Campi Flegrei, separating zones that present different geological and morphological features.

This partition is very useful to set the problem in a discrete framework, and the freedom in drawing the boundary of the zones allows an improved representation of the different geological and morphological features, including those offshore, that characterize each zone as well as the shape of the CI and NYT calderas that define the edge of the Campi Flegrei area and the spatial and temporal clustering of past vents. Apart from the areas between the CI and NYT calderas and the area offshore, where no past vents were located, the different zones had almost equivalent areal sizes so to avoid bias in the analysis. The spatial vent density for each zone was obtained by counting the number (or the fraction) of circles/ellipses of vent locations contained in the zone (Figure 2.4b). This alternative density distribution is consistent with the density contours obtained by kernel estimation and represents an a posteriori confirmation of the choice of the kernel bandwidth adopted. However, as expected, the computed peak values in the zones are now lower than those obtained with the kernel approach, because within each zone the spatial density is assumed uniform. Some of the main ideas adopted for the representation of the location of past vents will be relevant in the assessment of the model for vent clustering.

The information on vent distribution was integrated with an estimate of the number of lost vents in the three epochs. In several regions within the caldera and also outside it, several depositional units that cannot be correlated with identified vents have been recognized (*e.g.* [145]). Most of these deposits belong to events that occurred in Epoch I, which is why they are mostly buried below more recent sequences. The lost vents were assumed to be uniformly distributed over the on-land portion of the NYT caldera since no vent has been found outside this area, but alternative hypotheses about the location of these vents were also entertained, and their effects on final results investigated.

2.3.2. Distribution of faults and fractures

Faults and fractures represent the other two variables we used as input to the probability map of vent opening potential. Faults and fractures zones are in fact often correlated with the opening of new vents and typically represent a weakness element that may favor magma ascent and eruption. (*e.g.* [39, 24, 111]). However, the relationships between cropping out faults and fractures and localization of vent openings at Campi Flegrei and other volcanic settings is a complex issue still matter of debate. Several studies of rift zones show a close relationship between extensional regime on the Earth surface and ascending magma. Fields observations, geodetic and geophysical surveys, mathematical models and analogue experiments (*e.g.* [126, 108, 136, 101]) indicate that magma generally arises along vertical or steeply dipping dikes which produce subsidence on the Earth surface accommodated by normal faults and fractures. Moreover, the upward magma migration depends on several other features such as the dike-driving overpressure, density and viscosity of magma and the physical properties of the hosting rocks. Usually most of the models assume an isotropic and homogeneous upper crust, whereas anisotropic and non-homogeneous features, such as stress barriers and pre-existing faults, can also control the pathway of the ascending magma. A sharp change in the elastic properties of the crustal sub-horizontal layering and the occurrence of weak layers can produce: (i) the arrest of the vertical dike propagation (e.g. [75]) or (ii) the lateral magma migration (e.g. [105]). Similarly, the existence of moderately-dipping shallow inherited faults and fractures can deviate the vertical dike propagation (e.g. [70, 99]), whereas, on the contrary, major deeply rooted subvertical faults can be preferred paths (e.g. [105]).

In this study we took advantage of a recent investigation carried out by Vitale and Isaia [153], which described and reconstructed the age, distribution and nature of the different type of faults and fractures located in the Campi Flegrei caldera. As far as faults are concerned, for our purposes, we focused on the map showing the maximum displacement of faults located in a given cell. In this way we assigned a weight not only to the presence of faults but also to the degree of displacement associated with them. In fact, according to [153] and [98], most of mesoscale faults hosted in the caldera are almost vertical with displacements ranging from few centimeters to several meters. Many faults are located close to volcanic vents both in the central portion and along the rims of the caldera. The inversion of faults data in [153] indicates a prevalence of NNE-SSW/NE-SW extensions in the central portion of the caldera, suggesting that an extensional stress field persists since, at least, about 4.2 ka BP (Figure2.5a).

Given the wide range of displacement observed in the field (see [153]), we assigned weights using a (four level) logarithmic (base 10) scale ranging from sub-centimetric to metric lengths, thus assuming that displacements less than few mm are negligible. There were additional assumptions as follows. The value ascribed to each cell was the maximum displacement of the faults cutting the cell. In a case where no faults were recognized in a given cell due to the presence of overlying geological or anthropic structures, estimates were made using information from regional structures, morphological structures and lineaments of gravity anomalies (see [69, 28, 153]). The resulting map (Figure 2.5a) clearly shows that the larger values of displacement are inferred to be present in the central part of the caldera (e.g. Pozzuoli, Solfatara, San Vito) and along the eastern and western borders of the NYT caldera (see also Figure 2.1). This distribution was then normalized to obtain an integral sum equal to one, and the adjusted distribution was used in the defini-

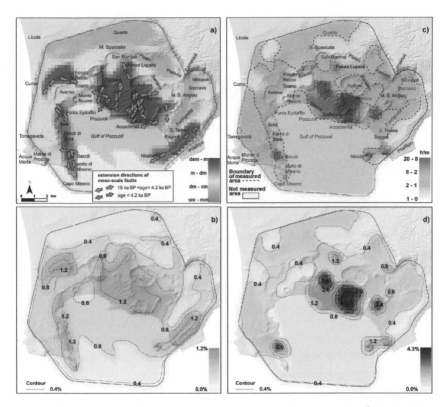

Figure 2.5. (a) Distribution of the maximum fault displacement in the caldera as derived from the dataset of [153]. The four colour levels shown correspond to displacements of different orders of magnitude ranging from sub-centimetric to metric scales. The figure also shows the extensional directions associated to the main mesoscale faults of the caldera with specific indication of those active in the last 4.2 ka. (b) Density distribution of the probability of vent opening obtained normalizing the values of maximum fault displacement. (c) Distribution of the surface fracture density in the caldera as derived from the dataset of [153]. In this case the four colours correspond to different values of density ranging between about 1 and 20 fractures per meter (fr/m). Wide areas of the on-land caldera and the offshore part were not measured (dashed areas). In these areas the average value of the total measured zone was assumed. (d) Density distribution of the probability of vent opening obtained normalizing the values of surface fracture density. In Figures 1.6b and 1.6d values indicate the percentage probability of vent opening per km^2 (conditional on the occurrence of an eruption). From [17].

tion of the vent opening probability map (see Figure 2.5b). The resulting map is more homogeneous than the probability maps based solely on the spatial distribution of known past vents, with maximum probability of about 1.2% per km^2 in the areas with greatest displacements, mentioned above.

Fractures were also elements assumed representative of past and present deformation within the caldera and therefore of potentially weakness areas likely exploited by magmatic fluids and indicative of future vent opening (*e.g.* [131, 99, 111]). However, the definition of surface fracture density is particularly challenging due to the sparse, diverse and incomplete nature of available measurements. The value assigned to each cell was derived by the number of surface fractures naked eye measured per meter of survey line length at each site (see [153] for the locations of the sites). In case there was a single measurement site in a cell we simply took that as the measured linear density value, whereas when there were multiple sites we conservatively assumed the highest measured linear density value for caution. Being the fracture density dependent on many factors, including the bed thickness, lithology and texture (*e.g.* [10, 77]), the maximum value of density calculated for the lithotype most favorable to fracturing was assumed. In the case where a cell did not contain any measurements, a bilinear interpolation on neighboring cells was assumed.

Nonetheless, the resulting map remained largely incomplete since sizeable areas of the caldera were not close to any measurement. In those parts where measurements were available, weights were assigned proportional to surface fracture density on a linear scale (Figure 2.5c). Values measured ranged from less than 1 up to about 20 fractures per meter. For areas with no data, a uniform value equal to the average value from the areas with measurements was assigned. The linear scale assumption was based on the fact that fracture openings show small variations, normally <1 mm and rarely larger than a few centimeters (see [153]). By assuming that fracture openings range between 0.2 and 1 mm (where 0.2 mm is the lowest opening threshold for naked-eye measurements) (*e.g.* [124, 77]), that fracture density ranges between 1 and 20 fractures per meter and that they are subvertical, the horizontal displacement ranges from 0.2 to 20 mm per meter.

The highest values of surface fracture density are located in the Solfatara area and around the town of Pozzuoli in the center of the caldera (Figure 2.5c). These areas correspond also to some regions of intense degassing and hydrothermal activity (see [34]). Other highly fractured areas are located at Averno, Bacoli-Capo Miseno, Nisida, Posillipo and part of the Agnano plain. As done for faults, the spatial distribution of fracture density was normalized to sum one across the whole caldera for using it as a component of the vent opening probability map (see Figure 2.5d): in this way the probability density is defined as directly proportional to the fracture density. Based on this assumption, the maximum percentage probability of vent opening per km^2 reaches values of about 4.3% in the very highly fractured zones, mentioned above, with peak values comparable to the maxima of spatial vent density computed from past events (see Figure 2.4).

2.4. Results

Once the five spatial density maps described above were constructed, *i.e.* the three distributions of vent location in the three epochs, the distribution of maximum fault dislocation and the distribution of the surface fracture density, we applied the structured expert elicitation techniques described in Appendix A of this chapter and in Chapter 5. As explained above, the maps related to faults and fractures, assimilated here into vent opening probability maps (see Figure 2.5b, d), are strictly maps of maximum fault dislocation and surface fracture density, respectively, this meaning that their contributions to the probability of vent opening are greater where net dislocations and densities are greater. An alternative uniform distribution over the whole caldera area was also adopted to represent the possibility there may be no correlation between the vent opening distribution and the five variables considered here.

Several elicitation sessions, involving about 8-10 experts with different volcanological backgrounds, were carried out during the three-year long study through meetings and remote consultations. The main goal was to achieve transparent, robust and shared estimates of the unknown values of target variables. We carried out the expert calibration by using seed (or test) questions on a mix of Campi Flegrei and Vesuvius volcanism and, more generally, on explosive volcanism (see Chapter 6).

The elicitation was based upon target questions that followed a hierarchical logic tree structure with various levels (see Appendix A of this chapter; Figure 2.8 and Table 2.1). First, the relevance of the five considered spatial variables was compared to that of the homogeneous distribution. At the next level, the contribution of past vent distributions was compared to those of structural features (*i.e.* fault displacement and surface fracture density). Moving to the next level, the relative importance of single vents in the three epochs was evaluated as well as the relative weight of the faults and fracture distributions. Additional target questions were related to the number of lost vents in each epoch, based on the stratigraphic evidence available.

2.4.1. The weights of the variables

Table 2.2 and Figure 2.6 illustrate, respectively, the percentiles and the density distributions of the weight of the five variables, derived from the

Questions/Percentiles	5%ile	50%ile	95%ile
Q1: how much weight do you assign to the information provided jointly by the distributions of the five variables considered here? (%)	50	80	95
	62	81	92
	46	83	95
Q2: how much weight should be given to a spatially uniform distribution? (%)	5	20	50
	8	19	37
	5	16	54
Q3: how much weight would you give to the distribution of the location of past vents (considering jointly all three epochs)? (%)	41	63	89
	48	64	81
	45	63	88
Q4: how much weight would you give to the distribution of the 'structural features' (considering jointly faults and fractures)? (%)	19	39	60
	23	36	52
	15	37	55
Q5: how much weight you would give to a single vent opened in the Epoch I (33 vents identified)? (%)	9	32	50
	14	28	43
	9	26	67
Q6: how much weight you would give to a single vent opened in the Epoch II (8 vents identified)? (%)	10	33	54
	16	34	47
	13	31	56
Q7: how much weight you would give to a single vent opened in the Epoch III (plus Monte Nuovo) (29 vents identified)? (%)	20	35	70
	27	40	62
	14	41	70

Questions/Percentiles	5%ile	50%ile	95%ile
Q8: how much weight do you give to the spatial distribution of faults? (%)	30	60	80
	36	59	67
	20	60	84
Q9: how much weight do you give to the spatial distribution of fractures? (%)	20	40	61
	24	41	62
	16	41	79
Q10: based on the information provided how many vents are missing in the dataset of vent positions of the Epoch I (33 vents recognized)?	5	7	10
	5	7	11
	4	7	13
Q11: based on the information provided how many vents are missing in the dataset of vent positions of the Epoch II (8 vents recognized)?	0	1	2
	0	1	2
	0	1	3
Q12: based on the information provided how many vents are missing in the dataset of vent positions of the Epoch III (plus Monte Nuovo) (29 vents recognized)?	1	2	4
	1	2	4
	0	2	5
Q13: considering the three epochs and based on the information provided, what percentage of CF eruptions on average do you think could have occurred simultaneously (i.e. within a few weeks of each other, or less) from two (or more?) different vents (i.e. located a few kilometers of each other)? (%)	5	10	24
	5	14	29
	3	11	73
Q14: considering the areas that have been invaded by PDCs from CF in the three epochs and based on the information provided, what would be the typical lateral underestimation error for the mapping/positioning of PDC invasion area perimeter boundaries (m)?	140	500	1000
	150	390	920
	110	360	1170

Table 2.1. Abbreviated target questions of the formal elicitation procedure, and probability percentages of the 5th, 50th and 95th percentiles of the DMs obtained. The three values reported in each cell refer, from top to bottom, to the CM, ERF and EW models. Because of the skewness of the probability distributions combined, some of the median values of complementary questions do not sum exactly to 100%.

elicitation procedure. The weights for lost vents and for the homogeneous map are also included. The weight for lost vents comes from the sum of the products of the number of lost vents in each epoch with the relative weight of a single vent (see Table 2.1). Elicitation outcomes are reported for the three different models, *i.e.* (a) the Classical Model (CM) of Cooke [41]; (b) the Expected Relative Frequency (ERF) model of Flandoli *et al.* [68] and (c) the Equal Weight (EW) model. It appears from the results that the outcomes from the three models are consistent with one another, overall, and do not show any gross differences. As expected the EW model produced wider uncertainties relative to the performance-based CM and the ERF model solutions. In the following we refer mostly to the CM model solutions since it is the most appropriate approach for capturing the uncertainty on unknown values of variables. Quite similar and slightly narrower distributions are computed by the ERF model which, typically, is more precise than the CM in estimating the central value of a distribution. Basic robustness tests show that the CM results are stable when the responses of different sub-groups of experts, determined in terms of specific expertise and background, are processed separately; this creates confidence that the elicitation process has reliably and validly synthesized the group's views on the scientific issues involved.

Variable/ Statistics	Vents Epoch I	Vents Epoch II	Vents Epoch III	Lost vents	Faults	Fractures	Homog. map
5%ile	6.3	1.3	10.2	3.3	8.1	5.4	6.3
	9.5	2.2	14.7	4.2	10.2	7	8.7
	6.3	1.5	7.6	3.4	5.3	4.3	6.5
Mean	16	4.5	20.4	5.9	16.4	11.9	24.9
	16.4	4.8	22.5	6.3	16.5	12.3	21.3
	17.7	4.6	19.3	6.7	13.0	12	25.9
95%ile	26.7	8.7	33.3	9	26.6	20.4	42.4
	24	7.6	31.6	8.8	23.9	18.6	31.1
	30.5	9.3	33.8	11	24.3	22	45.4

Table 2.2. Probability percentages of the mean and 5th and 95th percentiles of the weight of the five variables considered together with the weights of the lost vents and homogeneous map. The three values reported in each cell refer to, from top to bottom, the CM, ERF and EW models. The median values (*i.e.* the 50th percentile) are very similar to the mean values, within about 1%.

From Table 2.2 it emerges that the weight assigned to the distribution of the vents of Epoch III is the largest with a value of about 20% (*i.e.* mean value) and a credible interval between about 10% and 33% (corresponding to the 5th and 95th percentiles, respectively). Weights of about 4.5% and 16% were estimated for the mean of vent distributions of Epoch I and Epoch II, respectively, reflecting to some extent the much larger numbers of vents that occurred in Epoch I. The mean weight of lost vents was estimated at about 6%, with a credible interval between about 3% and 9%. In fact, it was estimated that between 5 and 10 vents were lost from the first epoch set, between 0 and 2 from the second epoch and between 1 and 4 from the third epoch (see Table 2.1). The distribution of fault displacement was weighted about 16.5% whereas that of fractures about 12% (mean values). Weights of faults and fractures were also af-

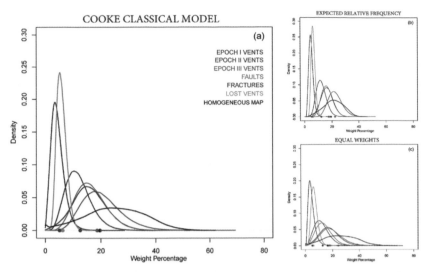

Figure 2.6. Density distribution of the weights of the six variables considered and of the lost vents as a function of the elicitation models assumed, *i.e.* , (a) Cooke CM, (b) ERF model, and (c) EW model. Along the x axis are also reported as coloured dots the estimates obtained by using just the best guess (central) values provided by the experts. Modified from [17].

fected by significant uncertainty with credible intervals ranging between about 8% and 27% for faults and between about 5% and 20% for fractures. Finally, a mean weight of about 25% was assigned to the homogeneous whole-caldera spatial distribution, with credible interval between about 6 and 42%.

Figure 2.6 represents the densities of the uncertainty spreads for the single weights as well as the central value weights obtained directly from the elicitation, represented as coloured dots along the x axis. Any discrepancies between these points and the mean values (reported in Table 2.2) depend on the skewness of the uncertainty distributions. These distributions show again that the CM provides marginally narrower probability density functions with respect to those from EW, and that ERF model distributions are still more narrow.

2.4.2. The maps of vent opening probability

Finally, Figure 2.7 shows the vent opening probability maps obtained from weighting and combining the six spatial distributions that were considered. The maps have been computed on the same 100x100 grid with cells of side 250 m used for the distribution of the five variables. The probability of vent opening is expressed as percentage of having a vent per km^2 conditional on the occurrence of a new eruption over the Campi

Flegrei caldera (so that the spatial integration of such probability map closes to 100%). The figure reports both the maps obtained using the kernel functions for the density distribution of past vents (Figure 2.7a, b and c) and the maps obtained using the partitioning of the caldera into sixteen homogeneous zones (Figure 2.7d, e, and f). Most importantly, thanks to the doubly stochastic structure of the model, the vent opening probability maps take into account the sources of epistemic uncertainty quantified here. This means that spatial vent opening probability is not depicted on a single map but through a set of maps which, in the figure, present mean values and 5th and 95th percentiles of the distributions associated with the density values of each cell (corresponding to Figures 2.7b, e, 2.7a, d and 2.7c, f, respectively). In particular, the spread between such percentiles is influenced by the uncertain correlation of the relevant variables with the position of opening of a new vent and by the uncertain number of past vents not included in the eruptive record; however, all the maps include uncertainties on the locations of past vents. The maps presented here refer to the estimates obtained using the CM solutions, and assume that the contribution of lost vents is uniformly distributed over the on-land portion of the NYT caldera. Similar maps, not reported here, have been produced by using the other weighting models and assuming alternative distributions for the lost vents (*e.g.* similar to those of the identified vents); however, they do not produce significantly different outcomes.

All the maps of Figure 2.7 show that the vent opening position probability is widely spread over the caldera. With specific reference to the mean probability maps (Figure 2.7b, e), it appears that the probability of vent opening per km^2 is, roughly speaking, greater than 0.4% over all the NYT caldera, with values below about 0.1% just in the area between the NYT and CI calderas, whereas the probability is about 0.2% in the portion of Collina di Posillipo examined. These latter values, in particular, derive from the contribution of the fault and fracture structures existing between the two calderas boundaries and on Collina di Posillipo (see Figure 2.5). No significant differences appear between the probability values on the map based on the kernel functions and those of the map based on the caldera partitioning.

From all the maps, the existence of a wide region of high probability of vent opening, located approximately in the area of Astroni-Agnano-Solfatara, also emerges. Probability values of vent per km^2 up to about 2.4% are predicted by the mean map, from both kernel-based and partition-based maps (Figure 2.7b, e). Credible intervals for these highest values range between about 1.6% and 3.2% (see Figure 2.7a, d and 2.7c, e). The zone of Pisani, north of Astroni, is also characterized by sig-

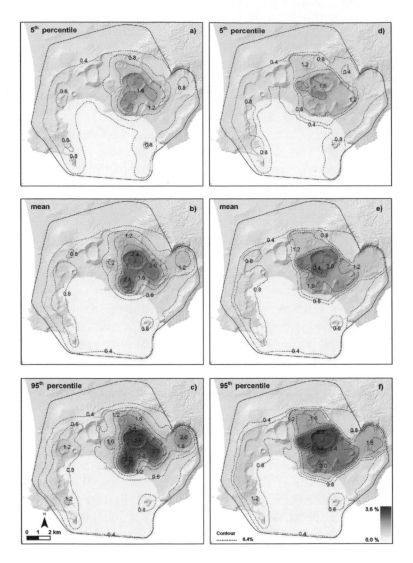

Figure 2.7. Probability maps of new vent opening as obtained weighting the six variable distributions considered. Contours and colours indicate the percentage probability of vent opening per km² (conditional on the occurrence of an eruption). (a-c) The use of kernel functions for the estimate of the density of past vents. (d-f) The partition of the caldera in the 16 homogeneous zones (see text for further explanations). Figures 1.8a and 1.8d refer to the 5th percentile, Figures 1.8b and 1.8e to the mean values, and Figures 1.8c and 1.8f to the 95th percentile. The median maps result very similar to the corresponding mean maps. From [17].

nificant values of about 1.2%. In this high probability region, too, the estimates obtained using the kernel functions are consistent with those obtained using the partitioning. As mentioned above, the main difference is that the kernel distributions concentrate the probability more in the centers of the clusters of past vents, whereas the partition approach distributes the probability uniformly over the single zones, based on broader volcanological and structural features. However, in both cases the highest probabilities were found in the Astroni area.

Outside this higher probability central area, probability of vent opening is quite dispersed with secondary maxima in the zone of Soccavo, in the eastern part of the caldera, and in the zones of Averno-Monte Nuovo-Baia Capo Miseno, in the western part. At Soccavo, vent opening probability values are about 1.2%; in the western part of the caldera they reach 1% (as mean values). The zones of Gauro-Toiano in the central-western part of the caldera appear to be associated with the lowest probability of vent opening; the offshore area is characterized by mean values of about 0.5%. In Table 2.3 we report the estimates of vent opening probability distribution integrated over each of the sixteen zones of the caldera, computed using the caldera partitioning and the kernel functions, and by adopting the CM, ERF and EW models.

2.5. Discussion

The forecast of the location of a future vent is a challenging goal of volcanology and an important element for volcanic hazard assessment. This is particularly true for calderas that typically show eruption behaviour patterns significantly more complex than central volcanoes. Most of the known calderas have produced eruption sequences which originated from significantly dispersed vents, difficult to associate into any regular time-space pattern. Moreover, in most cases explosive eruptions are prevalent and show remarkably variable scales of intensity and magnitude. The way a caldera evolves also favors the development of significantly complex structures, abundant hydrothermal circulation and thermal anomalies, all of which further complicate the problem and make forecasts particularly uncertain (*e.g.* [1]). Campi Flegrei presents many of the above-mentioned properties of calderas. The main characteristics of the Campi Flegrei system are: the presence of a large caldera, produced by two very large explosive eruptions (CI and NYT), with many smaller calderas and craters located inside it; repeated long periods of quiescence, lasting millennia, interrupted by periods of activity lasting several centuries (eruptive epochs); the prevalence of explosive eruptions, and the remarkable spatial scatter of vents active in the last 15 ka over the whole NYT caldera. This latter feature is particularly relevant for Campi Flegrei due to the significant size of the caldera (about 12 km diameter) and the dense urbanization of the territory. A vent opening map is therefore key to providing adequate hazard maps for the main explosive phenomena for which this volcano system is notable.

We have produced several vent opening probability maps based on the latest knowledge of the volcano history and quantifying some of the main epistemic uncertainties identified with this complex volcanic system. In addition to the consideration of the distribution of vents that occurred in the last 15 ka, including the contribution of lost vents, the analysis accounts for the potential influence of faults and surface fractures on the opening of future vents. We assume the presence of these features indicative of areas of upper crustal weakness in the caldera that could affect magma intrusion in a case of unrest. The structural survey of the caldera (see [153]) indicates that faults and fractures acted in various periods of the last 15 ka of caldera evolution, especially in the central area and along the caldera rims. In particular in the central area the youngest faults (dated from about 4.2 ka to present, see Figure 2.5a) show a common extensional stress field characterized by a NNE-SSW/N-S extension that has been interpreted as a favorable condition for possible future magma intrusions. Consideration was also given to the fact that caldera systems are particularly complex and vent opening could be also affected by other variables not accounted for in this analysis. Thus a uniformly distributed contribution of other influences was assumed over the whole caldera to represent the incompleteness of our knowledge and understanding of the system.

Through several meetings and open discussions, the study participants deliberated in depth on the volcanological datasets to be adopted, as well as their meaning for the specific purpose of this probabilistic analysis. Spreads of opinions were then evaluated and aggregated through structured expert elicitation (see Appendix A of this chapter), to represent and optimize group judgments. The findings of the analysis were revised through several iterations to fully refine and clarify the data considered and reach an acceptable consensus on outcomes. Findings were also evaluated by adopting alternative elicitation pooling models. The outcomes were found substantially robust with respect to the choice of the expert aggregation method (CM, ERF or EW), the statistical central value presented (median, mean, or mode of elicited values), and the approach used to produce the probability map based on past vent locations (kernel-based or caldera partitioning).

Based on the expert elicitation outcomes, location distributions of previous vents are judged the most important variables for quantifying the vent opening probability map, with a total contribution weight of about 47% (mean value). This estimate includes a weight of about 6% related to the lost vents. In detail, the location distribution of the vents of Epoch III receives the largest weight (about 20% as mean value) followed by the distribution of the vents of Epoch I and Epoch II with about 16% and 4.5% (as mean values), respectively. During the elicitation, experts were asked to assign weights to the individual vents of each epoch. Then the weight of the vent location distribution for each epoch was computed as the product of the weight of a single vent and the number of vents that occurred in that epoch. This is the reason for the larger relative weight of the location distribution of vents of Epoch I (33 vents) with respect to that of Epoch II (8 vents). The weights ascribed to a single vent in the three epochs at 32%, 33% and 35% for Epoch I, II and III, respectively (see Table 2.1), were remarkably similar and suggest the experts did not have any meaningful preference for data from one epoch over any other. The distribution of the maximum fault displacement and surface fracture density were weighted (as mean values) about 16% and 12%, respectively. This outcome is related to the fact that, for most of the group members, faults appeared more related than fractures to the deep system of the volcano and therefore possibly reflect potential regions of future vent opening. In fact, faults produce much larger deformations than fractures suggesting a closer link to deep processes. Conversely, fractures were considered more representative of the status of the shallow layers of the caldera. Last of all, about 25% weight was assigned to the uniformly homogeneous spatial opening map, *i.e.* to the possibility that the next vent could open anywhere inside the NYT caldera. Additionally, it is very important to note that, from the elicitation, all these weight estimates were characterized by significant associated uncertainties.

Our final maps (Figure 2.7), obtained by weighting and combining the five distributions described above plus the lost vents represented as a uniform distribution, provide a quantitative assessment of the spatial probability of vent opening within the caldera. The results highlight the existence of a main, quite wide, region in the central-eastern part of the caldera characterized by the highest probabilities of vent opening. This region corresponds to the area of Averno-Agnano-Solfatara. Although the detailed probability distribution in this area depends to some extent on the type of numerical approach used (*e.g.* kernel functions vs caldera partitioning), the maximum values of vent opening probability per km^2 are in the credible interval [1.6%, 3.2%] with a mean value of about 2.4% in both cases. By spatial integration, the total probability of the next vent opening in this area is about 30% (see also Table 2.3). Secondary maxima are obtained in the western part of the caldera, *i.e.* zones of

COOKE CLASSICAL MODEL

CALDERA PARTITION

	CMiseno	Baia	Averno	Gauro	Toiano	Solfatara	Pisani	Astroni	Agnano	Pianura	Soccavo	Bagnoli	Nisida	Mare	Cl-outer	Posillipo
5th percentile	4.3%	4.5%	4.1%	3.5%	3.0%	7.6%	5.9%	9.7%	8.1%	2.7%	4.3%	3.0%	3.7%	8.7%	2.6%	0.8%
mean	4.7%	5.4%	4.9%	4.4%	3.8%	9.2%	7.5%	13.1%	10.0%	3.3%	6.8%	3.7%	4.0%	13.1%	4.2%	1.2%
95th percentile	4.9%	6.5%	5.8%	5.2%	4.5%	11.1%	9.2%	17.3%	12.4%	3.9%	9.2%	4.3%	4.3%	17.1%	6.0%	1.8%
(5%ile-mean)/mean	-8.2%	-17.5%	-16.2%	-21.8%	-20.8%	-17.2%	-20.9%	-25.4%	-18.4%	-19.8%	-36.1%	-18.3%	-8.7%	-33.9%	-37.7%	-38.7%
(95%ile-mean)/mean	6.0%	19.3%	17.6%	17.7%	17.7%	20.9%	23.6%	32.8%	23.9%	17.1%	36.4%	15.3%	7.2%	30.4%	41.7%	42.7%
50th percentile	4.7%	5.4%	4.9%	4.5%	3.9%	9.1%	7.4%	12.8%	9.8%	3.4%	6.7%	3.8%	4.1%	13.2%	4.2%	1.2%
best guess	4.7%	5.7%	4.6%	4.6%	3.7%	8.9%	8.0%	12.6%	9.9%	3.3%	7.5%	3.7%	4.0%	12.2%	4.6%	1.4%

KERNEL DENSITY ESTIMATION

	CMiseno	Baia	Averno	Gauro	Toiano	Solfatara	Pisani	Astroni	Agnano	Pianura	Soccavo	Bagnoli	Nisida	Mare	Cl-outer	Posillipo
p05%	4.1%	4.0%	4.4%	3.4%	3.5%	7.3%	5.9%	9.2%	8.2%	3.9%	4.3%	3.2%	3.4%	9.7%	4.2%	1.0%
mean	4.5%	4.6%	4.9%	4.1%	4.2%	8.8%	6.7%	11.9%	9.9%	4.6%	6.4%	3.9%	3.8%	13.9%	5.8%	1.5%
p95%	4.8%	5.2%	5.5%	4.7%	4.8%	10.7%	7.7%	15.3%	12.1%	5.4%	8.5%	4.4%	4.2%	17.6%	7.6%	2.0%
(5%ile-mean)/mean	-8.7%	-13.9%	-11.4%	-16.9%	-16.3%	-17.0%	-12.2%	-22.7%	-17.2%	-15.6%	-32.7%	-16.6%	-10.7%	-30.0%	-28.4%	-32.2%
(95%ile-mean)/mean	6.7%	13.3%	11.8%	13.7%	14.2%	20.5%	14.3%	29.0%	22.3%	15.8%	32.8%	14.0%	9.4%	26.7%	30.9%	36.3%
p50%	4.5%	4.6%	4.9%	4.1%	4.2%	8.8%	6.7%	11.6%	9.7%	4.6%	6.4%	3.9%	3.8%	14.0%	5.8%	1.4%
best guess	4.5%	4.7%	4.7%	4.1%	4.0%	8.6%	7.0%	11.5%	9.8%	4.8%	7.0%	3.9%	3.7%	13.1%	6.5%	1.6%

EXPECTED RELATIVE FREQUENCY

CALDERA PARTITION

	CMiseno	Baia	Averno	Gauro	Toiano	Solfatara	Pisani	Astroni	Agnano	Pianura	Soccavo	Bagnoli	Nisida	Mare	Cl-outer	Posillipo
p05%	4.4%	4.7%	4.4%	3.6%	3.2%	8.3%	6.4%	11.2%	9.0%	2.8%	5.0%	3.1%	3.7%	9.2%	3.1%	0.9%
mean	4.6%	5.4%	5.0%	4.3%	3.7%	9.5%	7.5%	13.8%	10.3%	3.2%	6.7%	3.6%	4.0%	12.2%	4.3%	1.2%
p95%	4.8%	6.1%	5.6%	4.9%	4.2%	10.8%	8.7%	16.8%	12.0%	3.7%	8.5%	4.0%	4.2%	15.0%	5.5%	1.6%
(5%ile-mean)/mean	-5.8%	-13.0%	-11.8%	-15.6%	-14.8%	-12.7%	-15.1%	-18.5%	-13.4%	-14.2%	-26.0%	-13.3%	-6.0%	-24.8%	-27.0%	-27.4%
(95%ile-mean)/mean	4.8%	13.8%	12.6%	14.0%	13.7%	14.2%	16.5%	21.9%	16.0%	13.0%	26.0%	11.7%	5.3%	22.9%	28.4%	29.8%
p50%	4.6%	5.4%	5.0%	4.3%	3.7%	9.4%	7.5%	13.6%	10.3%	3.3%	6.7%	3.6%	4.0%	12.3%	4.3%	1.2%
best guess	4.6%	5.5%	4.9%	4.3%	3.6%	9.4%	7.7%	13.8%	10.5%	3.2%	6.9%	3.6%	3.9%	11.8%	4.4%	1.3%

KERNEL DENSITY ESTIMATION

	CMiseno	Baia	Averno	Gauro	Toiano	Solfatara	Pisani	Astroni	Agnano	Pianura	Soccavo	Bagnoli	Nisida	Mare	Cl-outer	Posillipo
p05%	4.1%	4.1%	4.5%	3.5%	3.6%	8.0%	6.2%	10.4%	8.9%	4.1%	4.9%	3.3%	3.5%	10.2%	4.8%	1.1%
mean	4.4%	4.5%	4.9%	4.0%	4.1%	9.1%	6.8%	12.4%	10.2%	4.6%	6.4%	3.8%	3.8%	13.1%	5.9%	1.5%
p95%	4.6%	5.0%	5.4%	4.4%	4.5%	10.4%	7.4%	14.9%	11.7%	5.1%	7.9%	4.2%	4.0%	15.7%	7.2%	1.8%
(5%ile-mean)/mean	-6.1%	-10.2%	-8.1%	-12.1%	-11.8%	-12.5%	-8.9%	-16.6%	-12.7%	-11.1%	-23.1%	-12.0%	-7.4%	-21.9%	-19.7%	-22.4%
(95%ile-mean)/mean	5.2%	9.7%	8.5%	10.8%	10.6%	14.0%	9.9%	19.4%	14.9%	11.4%	23.6%	10.7%	6.9%	20.2%	20.9%	25.2%
p50%	4.4%	4.5%	4.9%	4.0%	4.1%	9.1%	6.7%	12.3%	10.2%	4.6%	6.3%	3.8%	3.8%	13.1%	5.9%	1.5%
best guess	4.4%	4.6%	4.9%	4.0%	4.0%	9.1%	6.9%	12.5%	10.3%	4.6%	6.6%	3.7%	3.7%	12.6%	6.1%	1.5%

EQUAL WEIGHTS

CALDERA PARTITION

	CMiseno	Baia	Averno	Gauro	Toiano	Solfatara	Pisani	Astroni	Agnano	Pianura	Soccavo	Bagnoli	Nisida	Mare	Cl-outer	Posillipo
p05%	4.3%	4.6%	3.8%	3.4%	3.0%	7.1%	6.0%	9.0%	7.9%	2.7%	4.4%	3.0%	3.7%	8.2%	2.3%	0.7%
mean	4.7%	5.6%	4.8%	4.6%	3.8%	9.0%	7.8%	12.7%	9.9%	3.5%	7.3%	3.8%	4.0%	12.8%	4.0%	1.1%
p95%	5.1%	6.8%	5.8%	5.5%	4.5%	11.1%	9.9%	17.5%	12.6%	4.1%	10.2%	4.5%	4.4%	17.2%	5.7%	1.6%
(5%ile-mean)/mean	-9.3%	-18.4%	-20.5%	-25.1%	-21.8%	-20.9%	-23.2%	-29.5%	-20.3%	-21.8%	-39.7%	-20.6%	-8.9%	-36.2%	-40.8%	-41.1%
(95%ile-mean)/mean	7.2%	21.8%	20.7%	21.1%	18.7%	24.4%	26.9%	37.6%	26.9%	18.4%	41.2%	17.0%	8.2%	33.8%	43.5%	45.5%
p50%	4.7%	5.5%	4.8%	4.6%	3.8%	8.9%	7.7%	12.4%	9.7%	3.5%	7.2%	3.9%	4.1%	12.9%	3.9%	1.1%
best guess	4.6%	5.4%	5.0%	4.2%	3.6%	9.7%	7.6%	14.4%	10.7%	3.1%	6.7%	3.4%	3.9%	11.3%	4.6%	1.3%

KERNEL DENSITY ESTIMATION

	CMiseno	Baia	Averno	Gauro	Toiano	Solfatara	Pisani	Astroni	Agnano	Pianura	Soccavo	Bagnoli	Nisida	Mare	Cl-outer	Posillipo
p05%	4.1%	4.0%	4.2%	3.4%	3.4%	6.8%	6.0%	8.6%	7.9%	4.0%	4.4%	3.3%	3.4%	9.3%	3.9%	0.9%
mean	4.5%	4.7%	4.9%	4.2%	4.1%	8.6%	6.9%	11.6%	9.8%	4.8%	6.8%	4.0%	3.8%	13.6%	5.7%	1.4%
p95%	4.9%	5.4%	5.5%	4.8%	4.7%	10.7%	8.0%	15.5%	12.2%	5.7%	9.4%	4.6%	4.2%	17.7%	7.5%	1.9%
(5%ile-mean)/mean	-10.0%	-14.9%	-13.8%	-19.4%	-17.7%	-21.0%	-13.0%	-26.1%	-19.2%	-16.6%	-35.5%	-18.1%	-11.2%	-32.0%	-31.0%	-34.6%
(95%ile-mean)/mean	7.7%	15.4%	13.6%	15.8%	14.6%	24.1%	15.9%	33.0%	24.6%	17.4%	37.1%	15.6%	10.2%	29.9%	33.3%	39.0%
p50%	4.5%	4.7%	4.9%	4.2%	4.2%	8.6%	6.8%	11.4%	9.6%	4.8%	6.8%	4.0%	3.9%	13.7%	5.6%	1.3%
best guess	4.3%	4.4%	4.9%	3.9%	3.9%	9.4%	6.8%	13.0%	10.6%	4.5%	6.4%	3.6%	3.7%	12.2%	6.3%	1.6%

Table 2.3. Integrated probabilities of vent opening in the sixteen zones of our caldera partition (see Figure 2.4b) by using the three scoring models CM, ERF and EW and the two approaches based on the caldera partitioning and the kernel density estimation (Figure 2.7). Central values (*i.e.* mean, median, and the best-guess/modal values given by the experts), 5th and 95th uncertainty percentiles of such integrated probabilities, and also the uncertainty ranges with respect to the mean value, are reported.

Averno-Monte Nuovo-Baia-Capo Miseno, and in the Soccavo and Pisani areas. These areas are characterized by mean values of about 1-1.2%, *i.e.* less than half the values estimated for the highest probability area. However, the probability of vent opening is not confined to these areas and, with mean values everywhere above 0.4%, the possibility is widespread over the caldera, thus making the associated background hazard potential broadly distributed in space. It is worth noting also that the last eruption of Campi Flegrei (Monte Nuovo) occurred in one of the secondary max-

ima of the computed vent opening probability map and not in the area
with the highest probability.

Contour	Area (km^2) [5th - mean - 95th percentiles]	Percentage of CF area [5th - mean - 95th percentiles]
Whole CF	144	100%
0.5% prob./km^2	65 - 81 - 107	45% - 56% - 74%
1% prob./km^2	23 - 31 - 36	16% - 21% - 25%
2% prob./km^2	0.2 - 5 - 12	0.1% - 3% - 8%

Table 2.4. Extension of the principal areal vent opening probability contours
(see Figure 1.8d, e, f): areas (in km^2) and percentages of the whole Campi
Flegrei area, with associated uncertainties.

Table 2.4 reports the total areas of the main vent opening probability
contours, the proportions of the whole caldera they occupy, expressed as
percentages, and the associated uncertainties on these spatial parameters.
More than half the caldera has an average probability of vent opening
greater than 0.5% and more than one-fifth has a probability larger than
1% per km^2. Furthermore, the quantified area uncertainty estimates that
we provide on these contoured areas should prove valuable when considering confidence levels in mitigation decisions.

It is worth highlighting also that the vent opening probability values per
km^2 are associated to a substantial uncertainty range, here represented
as $(5^{th}\ perc - mean)/mean$ and $(95^{th}\ perc - mean)/mean$; based
on inspection of data, it is on average about ±30% of the mean value,
with variations from ±10% to ±50% (corresponding to 5th and 95th percentiles) in different areas of the caldera. In particular, uncertainty values
spatially change as a function of the variables considered and the way
their uncertainties vary and influence the aggregated weights. Estimates
of the uncertainty range of integrated probabilities of vent opening on the
zones of the caldera partition are also reported in Table 2.2.

From inspection of numerical outcomes, the probability of vent opening in the offshore portion of the caldera is about 25±5%. This is a significant value suggesting further investigation of such a possibility would be
worthwhile. Also, our knowledge and hence inferences about this portion
of the caldera are affected by the lack of information compared to the on-
land areas (see Figure 2.5 about faults and fracture distributions). From
the maps created it is also possible to estimate that the mean probability
of vent opening in the eastern part of the caldera (assuming the division
line between the western and eastern parts coincides with the N-S bor-

der dividing the zones of Gauro-Toiano, on the west, from the zones of Pisani-Astroni-Solfatara, on the east) is significantly larger than that in the western part (66% vs 34%, with an uncertainty around ±4%).

Our present results appear qualitatively consistent with those of [3] and [122], which suggested the areas in the central-eastern part of the caldera are those with the greatest likelihoods of vent opening. They are also semi-quantitatively consistent with those of Selva *et al.* [141] although, relative to them, our study indicates the area with the highest likelihood of vent opening (*i.e.* Astroni-Agnano-Solfatara) has substantially greater probability values than in other parts of the caldera. With reference to mean probability values, the area of Astroni-Agnano-Solfatara has a maximum of 2.4% per km^2, against the 1.9% reported by [141]. Conversely, in the western portion of the caldera the high mean values are about 1% as against 1.5% of [141] In other words, in this study the probability of vent opening appears substantially more concentrated in the central-eastern part of the caldera with respect to the estimates of [141], which indicate a main secondary maximum in the western part of the caldera.

Further, [141] report uncertainty ranges of the vent opening probability per km^2 that appear significantly greater than those reported in this study, over the whole caldera: their average spread values correspond to about +180%/-90% of the local mean values, compared with our estimate of ±30%. This is probably due, in part at least, to their assumption of Dirichlet distributions defined with single global dispersion parameters. In contrast, the procedure adopted in our study allows uncertainties to be spatially modulated as a function of the different variables considered.

2.6. Appendix A: The expert elicitation technique

As discussed in the text, in this study we assumed that the background probability map of new vent opening can be expressed as a linear combination of the five density maps representing the spatial distributions of past vents (three maps, one for each of the three epochs), maximum fault displacement and surface fracture density, plus a homogeneous map that distributes the probability uniformly over the NYT caldera and a uniformly distributed contribute from lost vents on-land. The weights are affected by a significant amount of uncertainty representing unknown correlations of these relevant variables with the position of opening of a new vent. To estimate them and their variability we followed a structured elicitation procedure aimed at acquiring and understanding the experts' opinions. The same technique was used to quantify other relevant unknown variable values, such as the number of lost vents as well as other unknown values for variables used in the mapping of PDC hazard (see in the sequel).

2.6.1. Expert scoring rules and weighting assessments

The concept of expert elicitation concerns the adoption of a formal technique, or techniques, to be used to pool the judgments of a group of experts in order to inform decisions, forecasts or predictions based on a formalized treatment of uncertainties in relation to the matter under consideration (*e.g.* [41, 7], Chapter 5). In this particular study, the objective was not only the elicitation of the unknown value for a variable, but also its uncertainty properties. To this aim we applied three alternative expert scoring/weighting assessment models and we compared the different results obtained. The different expert scores are reported in Table 2.5.

Expert / EJ method	Expert 1	Expert 2	Expert 3	Expert 4	Expert 5	Expert 6	Expert 7	Expert 8
Classical Method	3.2%	0.0%	0.0%	0.0%	6.8%	0.0%	89.9%	0.1%
Expected Relative Frequency	11.9%	7.8%	7.3%	8.8%	19.4%	13.4%	20.9%	10.5%
Equal weights	12.5%	12.5%	12.5%	12.5%	12.5%	12.5%	12.5%	12.5%

Table 2.5. Expert scoring percentages (non including DM), assuming the different scoring rules. We remark that CM and EW follow the linear pooling, whereas ERF the quantile pooling (see in Chapter 5).

In general for a pooling or scoring scheme, the unknown variable values, for which estimates are needed, are called target items. During the elicitation procedure, each expert provided three values for every item: their judgment of the central value (represented by the median value of the uncertainty profile for Classical Model, or by the best guess with Expected Relative Frequency model) and then interval bounds which express his/her uncertainty about the credible range for the value. In our particular case the 5th and 95th percentiles of the uncertainty distribution were used as marker bounds for the uncertainty distributions. One way to aggregate the answers of a group collectively is that of calculating an equal weights pooling of the experts' densities: such a model is called the Equal Weights (EW) solution. This is the first of the three alternative pooling schemes that we considered.

However, this way of expressing a group opinion is often not optimal in terms of statistical informativeness: uncertainties tend to be very wide. To estimate value uncertainties accurately and informatively, empirical performance control (*e.g.* [41, 7]) is needed. With the Classical Model (see [41]), each expert is assigned a weight determined objectively on his/her ability to judge uncertainties with statistical accuracy and informativeness, thus providing a rational basis for pooling the views of a group of experts. In the Classical Model (CM), this empirical control is based on a set of seed questions. While actual values of these questions are retrievable, from the literature or other sources, experts are not expected to know them precisely but are expected to be able to define credible ranges that capture the values by informed reasoning. In our case the seed questions (reported in Chapter 6) were about carefully researched aspects of Campi Flegrei volcanism, other Italian volcanoes, such as Vesuvius, and about explosive volcanism in general. Experts' weights were then computed using the mathematical scoring rule process described in [41], with the resulting combination of experts' assessments on each target item referred to as a 'Decision Maker' (DM). Here we will briefly compare and contrast the Classical Model with a complementary approach, the Expected Relative Frequency model (ERF) (see [68]), and note these methods are based on similar but crucially different scoring rule philosophies.

The CM quantifies an expert's score as the product of two empirically determined measures, calibration and informativeness. The calibration score rewards good ability in an expert to be statistically accurate when assigning values to probability outcomes against known values. Thus, a 'well calibrated' expert provides answers such that the real values are symmetrically balanced with respect to his/her 50th percentile markers, and the majority fall between his/her 5th and 95th percentiles (but not necessarily all). The information score reflects an expert's capacity to provide concentrated distributions over the same variables. On its own, the information score does not consider the expert's distribution locations relative to the true realization values; instead, it is the relative information with respect to a uniform distribution, averaged on all the seed questions (global weights) or alternatively calculated and compared separately for each question (item weights). In summary, the CM takes the product of calibration and informativeness to reward jointly an expert with statistical accuracy and informativeness; we remark that the result is far more sensitive to the first one, and the CM score drops to zero rapidly if the calibration decreases. Furthermore, the CM may also perform optimization to maximize the resulting group 'decision maker' score (treating the DM as an added 'expert'). In particular it is implemented a cutoff threshold on the calibration score, that sets to zero the scores of the less calibrated experts if their presence decreases the performance score of the DM. This method tends to include in the DM only the 1-2 most calibrated experts, plus few others with very low scores. More details are also reported in Chapter 6.

The Expected Relative Frequency model computes a score for each seed item (*i.e.* the same calibration questions used in the CM) by integrating the probability density function over an interval centered around the true value. The idea of this model is that the expert's score is high if his/her mode is close to the true value (in relative error), but the score is modulated by the uncertainty declared by the expert. The average of an expert's scores over all the seed questions can be interpreted as the 'expected accuracy' of the expert (see [68]). In particular, if for each seed question a random variable is defined with the uncertainty distribution of a chosen expert, then his/her ERF score is the expectation of the fraction of these random variables that fall in a selected interval around the true values. The ERF tends to include each expert in the DM, even modulating his/her score in base of the performance, but there is another difference with respect to the CM that contributes to avoid the very large uncertainty ranges of the EW. It is implemented a different algorithm for combining the responses of the experts: instead of the linear pooling it is adopted the quantile pooling, with the effect of averaging the central values and reducing the uncertainty bounds (see in Chapter 5).

Relative to Equal Weights, and each other, in general the CM provides better quantification of variable uncertainty for multiple items, whereas the ERF model can provide more reliable estimates for central values. More information on the differences between the three models and their performance can be found in [68].

2.6.2. Logic tree questionnaire, linear weights and elicitation outcomes

To simplify the quantification of the weight to assign to each spatial distribution for the definition of the vent opening probability map, we defined a simple hierarchical logic tree (see Figure 2.8). Most of the target questions that were asked quantify the relative importance, or relevance, of one variable or feature of the system versus others. In each of these comparisons, the best estimate (central) percentages should sum close to 100%; strictly, the distribution means should sum to 100%, and the sum of elicited medians may diverge slightly depending upon tail asymmetries; elicited distributions can be normalized if necessary. Experts were asked to evaluate the uncertainties associated with their judgments of relative importance. The first question was about the relative importance of the five different distributions considered jointly compared to that of the uniform distribution, here assumed to represent the lack of information. At the next hierarchical level, the contribution of the overall past vent distribution was compared to that of the current structural features of the caldera (*i.e.* fault maximum displacement and surface fracture density). At the next level of the tree, the relative weights of single vents of the three epochs and the relative weights of faults and fracture distributions were evaluated. Other relevant questions were about the estimation of the number of lost vents in each epoch of activity, due to the successive eruptions and the morphological, volcanological and man-made transformations of the caldera, and to other variables with unknown values which might be relevant for the mapping of PDC hazard. We reported in Table 2.1 the abbreviated questionnaire protocol, and the corresponding CM, ERF and EW outcomes.

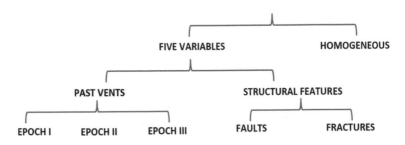

Figure 2.8. Hierarchical logic tree structure associated to the target questions queried during the elicitation sessions (from [17]).

To calculate the weights and their uncertainty characteristics for each of the three elicitation methods adopted, the three percentiles of the CM global DM were used for defining triangular distributions, from which random values could be sampled for each question. We followed a Monte Carlo simulation approach for determining these single branch weight estimates, normalizing complementary values to sum to one, and then multiplying the single weights over each branch of the logic tree. In this way we obtained a large sample of randomized weights to assign to the seven maps of the relevant spatial distributions, *i.e.* the distribution of past vents from the three epochs, the distribution of lost vents, assumed uniform, the fault and fracture distributions, and the uniform homogenous map over the NYT caldera (see Table 2.2 and Figure 2.6).

Each vector sample of the weights of the distributions (along with a sample of the number of lost vents of each epoch) can therefore be convolved into a probability map of new vent opening, obtained using those weights. To visualize the variability of these maps, we computed 'average maps' by Monte Carlo simulation, plotted as mean or median maps, as appropriate, and two maps representing the 'uncertainty bounds' of the distribution, expressed as the 5th and 95th percentiles of the values sampled (see Figure 2.7).

Finally, Table 2.3 reported the integrated probabilities on the zones of our caldera partition (see Definition 2.7), using the three scoring models

CM, ERF and EW and the two approaches based on caldera partitioning and on the kernel density estimation. The outcomes allow comparison of findings from the elicitation methods and from the density estimation methods. From inspection, it appears that the results overall are sensibly consistent with each other and that average discrepancies of mean values are generally below about 1%. Similarly, the upper and lower percentiles expressing the uncertainty bounds are substantially consistent between the different models. These outcomes confirm again that the CM and the ERF models produce narrower uncertainty distributions than the EW model. It is worth noting also that the distribution mean values and central values (medians or best estimates obtained directly from the elicitation) are remarkably similar to one another, and some significant differences arise only with the EW solutions.

2.7. Appendix B:
Dirichlet distributions and Bayesian inference

Let $\mathcal{P} = (A_l)_{l=1,...,N}$ be a finite partition of the Campi Flegrei caldera in $n \in \mathbb{N}$ parts; then a vent opening random variable X could be defined just on $\mathcal{P}$. A sample with respect to the uncertainty on such discrete vent opening map may be represented with the array θ of the probabilities of its n elements. Following this semi-parametric assumption the uncertainty distribution is directly defined on the $(n - 1)$-dimensional simplex $\mathcal{S}^{n-1} := \{\theta \in \mathbb{R}^n : \sum_i \theta_i = 1, \; \theta_i > 0 \; \forall i\}$ and a natural class of probability measures on that space is constituted by the Dirichlet distributions, that also possess useful properties in Bayesian statistics.

Definition 2.8 (Dirichlet distribution). Let θ be a random variable on $\mathcal{S}^n$ with probability density function

$$f_D(\theta) = \frac{\Gamma(\sum_i \alpha_i)}{\prod_i \Gamma(\alpha_i)} \prod_i \theta_i^{\alpha_i - 1},$$

where α is a vector of positive real numbers and here Γ is the Euler Gamma function. Then its probability distribution belongs to the Dirichlet class: $\theta \sim \mathcal{D}_n(\alpha_1, \ldots, \alpha_n)$.

In addition, Dirichlet distributions are easy to sample, thanks to the well known result that if $\theta \sim \mathcal{D}_n(\alpha)$ then $\forall i \leq n$

$$\theta_i = \frac{\zeta_i}{\sum \zeta_i},$$

where $(\zeta_i)_{i=1,...,n}$ is a family of independent random variables each of distribution $\Gamma(\alpha_i, 1)$. In the vent opening problem there are two different approaches that have been followed with Dirichlet distributions: the first was to take $\mathcal{P}$ as the family of the cells of a grid on the Campi Flegrei (see [141], where was $n = 700$), the second was to take $\mathcal{P}$ as a partition of the caldera in zones that are believed to have relevant differences, like in Definition 2.7 based on geological information. Another possibility is the implementation of the logic tree within a Bayesian uploading model, adopting Dirichlet distributions at each level of the tree. We remark that the main purpose of this chapter was the uncertainty quantification: the model presented is not Bayesian, but the long-term uncertainty distributions produced could be the basis also for the generation of up-dated short-term vent opening probability maps, also in a Bayesian framework. In particular the class of Dirichlet distributions is conjugate with the class of multinomial distributions by the Bayes rule.

Remark 2.9 (Bayes rule). Bayesian inference is widely adopted in assessing uncertainty problems: a priori uncertainty distribution is modified to a posteriori uncertainty by the observation of events. Either past events for long-term models or future events in case of real time changing models during crises. The likelihood of observed data A given a distribution vector (or a simpler parameter) θ is expressed by $P(A|\theta)$. The classical Bayes formula (one of its possible formulations) reads

$$f_{\text{post}}(\theta) = f_{\text{prior}}(\theta) \frac{P(A|\theta)}{P(A)}.$$

The multinomial distributions generalize the class of binomial distributions; an example of a multinomial random variable $Y \sim \mathcal{M}_m(y, p)$ is the array containing the number of occurrences of each of the values of a dice of n faces rolled m times, assuming p_i as the probability of the face i. In our case the variable Y may represent the number of past eruptions observed in each of the elements of $\mathcal{P}$ (in any combination).

Definition 2.10 (Multinomial distribution). Let Y be a random variable on $\mathbb{N}^n$ with discrete distribution

$$P(y) = \begin{cases} \frac{n!}{\prod_j^k (y_j!)} \prod_i p_i^{y_i} & \text{if } \sum y_i = m, \\ 0 & \text{otherwise,} \end{cases}$$

where $p \in \mathcal{S}^{n-1}$. Then its probability distribution belongs to the Multinomial class: $Y \sim \mathcal{M}_m(p)$.

In the following is the well known conjugation result of Dirichlet and Multinomial distributions.

Theorem 2.11 (Conjugation result). *Let $Z = (X^j)_{j \leq m}$ be a family of independent identically distributed variables on $\{1, \ldots, n\}$ whose discrete distribution is $\theta = (\theta_1, \ldots, \theta_k)$. For each given $z \in \{1, \ldots, n\}^m$ define $y(z) \in \{1, \ldots, m\}^n$ with $y(z)_i$ as the number of occurrences of the value i in the array z. Then the random variable $Y := y[Z]$ satisfies*

$$Y | \theta \sim \mathcal{M}_m(\theta).$$

Moreover assume θ is a random variable whose (a priori) distribution belongs to the Dirichlet class and that it is conditioned to an event $A = \{Y = y\}$, then $\theta | A = \theta | y$ (a posteriori) distribution still belongs to the Dirichlet class and

$$\theta \sim \mathcal{D}_n(\alpha) \quad \Rightarrow \quad \theta | y \sim \mathcal{D}_n(\alpha + y).$$

Even when the information available about past events is weaker and consists in the observation of an event in an unknown element of a subset of the partition $\mathcal{P}$, there is a generalized conjugation result about Dirichlet distributions mixtures.

Remark 2.12 (Dirichlet mixtures). Let be

$$C_\alpha = \Gamma \left(\sum_{i=1}^n \alpha_i \right) \bigg/ \prod_{i=1}^n \Gamma(\alpha_i)$$

the normalization constant of a Dirichlet distribution with concentration parameters array α. In particular it can be expressed as

$$C_\alpha = \left(\left(\sum_{i=1}^n \alpha_i \right) - 1 \right)! \bigg/ \prod_{i=1}^n (\alpha_i - 1)!$$

when α_i are positive integers (but this expression could be interpolated for any other positive value). It is

$$P(X = k) = C_\alpha \int_{Spl_{N-1}} \theta_k \prod_{i=1}^n \theta_i^{\alpha_i - 1} d\theta = C_\alpha / C_{\alpha + e_k}$$

where e_k is the k-element of the canonical base of $\mathbb{R}^n$. Hence it is easy to obtain that

$$f(\theta | X \in B) = \sum_{k \in B} \pi_k f(\theta | X = k),$$

where

$$\pi_k = \frac{1/C_{\alpha + e_k}}{\sum_{k \in B} 1/C_{\alpha + e_k}}.$$

It has been stated that the posterior of a Dirichlet, conditional on an event of the form $X \in B$ is a mixture of Dirichlets. Moreover, iterating the expression it is possible to prove that even if the prior distribution was a mixture of Dirichlet the posterior remains in that class, and the linear constants are easy to calculate.

The concentration parameters array α which rules a Dirichlet distribution is often represented through the array of average values Θ and the equivalent number of data Λ. This is a real number that represent a measure of the concentration of the distribution around its mean value; for each $i \leq n$ it is

$$\alpha_i = \Theta_i(\Lambda + n - 1).$$

It is called equivalent number of data because whenever an event $\{Z = z\}$ involving m samples is considered by the Bayes rule, the value of Λ increases of m (it easy to see from the definition, reminding that $\sum \Theta_i = 1$ either a priori and a posteriori, and that $\sum \alpha_i$ increases of m). Moreover it is well known that if $\Lambda = 1$ and $\alpha_i = 1/n$ for each $i \leq n$, then the correspondent Dirichlet distribution is uniform on S^{n-1}; a higher $\Lambda > 1$ implies a more than uniform concentration while a lower $\Lambda < 1$ produces dispersion from the mean values. But the adoption of Dirichlet distributions, so handy with Bayesian inference, has also some difficulties. Once is fixed the array of average values Θ, the parameter Λ rules all the model uncertainty, and the difference between the single locations depends only on the mathematical structure of each Dirichlet distribution, without any other degree of freedom: it is not possible to increase the uncertainty only in one place. Moreover the expression of the variance is

$$E\left[|\theta_i - \Theta_i|^2\right] = \frac{\Theta_i(1 - \Theta_i)}{\Lambda + n},$$

and it is easy to see that the more n is big, the less influence has the same increase of Λ on the value of the variance: in the limit of a very big n, the uncertainty ranges a priori and a posteriori remain almost the same. In addition, the adoption of a partition $\mathcal{P}$ constituted by a large number of small cells may produce some scattering phenomena on the samples: this is due to the impossibility to implement a major spatial correlation between closer cells in this kind of models. In conclusion it seems that with a simple Dirichlet distribution is not easy to capture the structure of the uncertainty distribution. Moreover, for developing a method able to include short-term information from the monitoring network, another difficulty is that the observed events do not coincide to an eruption in a particular location, that is the random variable of interest, but are only precursors of it and of hard interpretation.

Chapter 3
Pyroclastic density current invasion maps

3.1. Summary

Campi Flegrei is an active caldera containing densely populated settlements at very high risk of pyroclastic density currents (PDCs). We present here an innovative method for assessing background spatial PDC hazard with probabilistic invasion maps conditional on the occurrence of an explosive event. The method encompasses the probabilistic assessment of potential vent opening positions, derived in the previous chapter, combined with inferences about the spatial density distribution of PDC invasion areas from a simplified flow model, informed by reconstruction of deposits from eruptions in the last 15 ka. The flow model describes the PDC kinematics and accounts for main effects of topography on flow propagation. Structured expert elicitation is used to incorporate certain sources of epistemic uncertainty, and a Monte Carlo approach is adopted to produce a set of probabilistic hazard maps for the whole Campi Flegrei area.

Our findings show that, in case of an explosive eruption, almost the entire caldera is exposed to invasion with a mean probability of at least 5%, with peaks greater than 50% in some central areas. Some areas outside the caldera are also exposed to this danger, with mean probabilities of invasion of the order of 5-10%. Our analysis suggests that these probability estimates have location-specific uncertainties which can be substantial. The results prove to be robust with respect to alternative elicitation models and allow the influence on hazard mapping of different sources of uncertainty, and of theoretical and numerical assumptions, to be quantified. As in the other chapters, we base our probability model on the abstract definition of several random variables associated the volcanological phenomena of interest. These variables are then assessed assuming for them a doubly stochastic structure. The definition of the probability measure representing the area invaded by the next PDC phenomenon and its explicit construction separating epistemic uncertainty from physical variability as we did for the map of vent opening, is of the main importance in this chapter.

Definition 3.1 (The distribution of PDC invaded areas). Let Y be a random variable from the sample space $(\Omega, \mathcal{F}, P)$ to $(\mathbb{R}_+, \mathcal{B}(\mathbb{R}_+))$, representing the area invaded by PDCs during the next explosive eruption at Campi Flegrei. Let ν_Y be the probability measure that is the law of Y on $(\mathbb{R}_+, \mathcal{B}(\mathbb{R}_+))$; it is called a distribution of PDC invaded areas.

To construct a doubly stochastic model for this variable, we follow again the ideas and notation of Definition 1.1, projecting the space $(W, \mathcal{W}, M)$ on $(\mathbb{R}_+, \mathcal{B}(\mathbb{R}_+))$.

Definition 3.2 (Conditional distribution of PDC invaded areas). Let π^2 be a measurable function from $(W, \mathcal{W}, M)$ to $(\mathbb{R}_+, \mathcal{B}(\mathbb{R}_+))$, representing the projection of the physical space onto the eruptive scale space. We assume that

$$Y(\omega) = \pi^2 \left(\chi(\xi(\omega), \omega) \right), \text{ for almost every } \omega \in \Omega,$$

and we define the random variable $\check{Y}$ from $(E \times \Omega, \mathcal{E} \otimes \mathcal{F}, \eta \otimes P)$ to $(A, \mathcal{B}(A))$ as

$$\check{Y}(e, \omega) := \pi^2 \left(\chi(e, \omega) \right).$$

For each $e \in E$ the random variable $\check{Y}(e, \cdot)$ on $(\Omega, \mathcal{F}, P)$ represents the area invaded by the next PDC phenomenon at Campi Flegrei once adopted the epistemic assumption e. Its law $\nu_{\check{Y}}(e)$ is called probability distribution of PDC invaded areas conditional on the epistemic assumption e.

The definition of a very simple model representing the propagation of a PDC of a particular scale, once assumed its eruptive vent location is the second fundamental step.

Definition 3.3 (The simplified flow model). Let $B \subseteq \mathbb{R}^2$ be a domain such that $B \supseteq A$, representing an enlarged zone possibly affected by PDC hazard, considered with its Borel sigma algebra $\mathcal{B}(B)$. Let F be a functional from $A \times \mathbb{R}_+$ to the Borel subsets of B, such that $F(x, y)$ represents the set invaded by a PDC propagating from a vent x with a scale y.

Combining the map of vent opening μ_X (see previous chapter) with the distribution of PDC invaded areas ν_Y and by using the simplified propagation model F, it is possible to produce probabilistic hazard maps of PDC invasion and to give an estimate for their considered epistemic uncertainty.

Definition 3.4 (The maps of PDC invasion probability). Let p be a measurable function from $(B, \mathcal{B}(B))$ to $([0, 1], \mathcal{B}([0, 1]))$, that is defined as

$$p := E[1_{F(X,Y)}]$$

and represents the probability of each point of B to be reached by the next PDC. For each $z \in B$ we also define a random variable $\check{p}$ from $(E, \mathcal{E}, \eta)$ to $([0, 1], \mathcal{B}([0, 1]))$ as

$$[\check{p}(z)](e) := E[1_{F(\check{X}(e,\cdot), \check{Y}(e,\cdot))}](z).$$

It expresses the probability of each point of B to be reached by the next PDC, conditional on the epistemic assumption e.

In other words with $\check{p}$ we estimate the random probability of each point of B to be reached by the next PDC as a function of $e \in E$. This is calculated by a double Monte Carlo simulation with a nested structure.

Remark 3.5. Basic mapping of PDC hazard at Campi Flegrei has been already reported in previous studies. Some related to field reconstruction and numerical modelling of specific past events, while others endeavoured to produce specific or integrated PDC hazard maps in which variabilities of important parameters of the volcanic system, such as the eruption scale and vent location, were explicitly accounted for. For instance, Lirer *et al.* [100] reconstructed the distribution of PDC deposits from the main events of the last 5 ka and outlined a zonation of areas potentially affected by PDCs. Similarly, Orsi *et al.* [122], used field data to reconstruct the distribution of deposits in the last 15 ka and proposed a qualitative PDC hazard invasion map based on the last 5 ka of activity. In both studies, the eastern part of the caldera was found to have the greatest hazard exposure; however, the area considered for PDCs was limited to that within the caldera rim (*i.e.* excluding the Collina di Posillipo). Rossano *et al.* [134] proposed a hazard map based on a dynamic 1D Bingham flow model, which considered the variability of eruptive scale (including very large caldera-collapse events) and assumed a uniform vent opening probability in an area centered on the town of Pozzuoli. Todesco *et al.* [150] and Esposti Ongaro *et al.* [65], using 2D and 3D numerical multi-phase flow simulations of Plinian type events, analyzed in more detail the propagation dynamics of PDCs within the caldera to improve the description of the complex interaction between flows and topography. Those studies were focused specifically on the eastern sector of the caldera and showed that, for some positions of the vent, pyroclastic flows could overtop Collina di Posillipo, a notable topographical barrier for the central part of the city of Naples.

More recently, Alberico *et al.* [4], starting from the probability distribution of new vent opening positions of [3] and the occurrence probabilities of the three eruption size categories of the last 5 ka, as defined by Orsi *et al.* [123], produced a qualitative integrated hazard map of PDC invasion for the city of Naples with five levels of hazard (Vesuvius hazards were also included in the map). Similar maps, this time associated with various eruption VEIs (Volcanic Explosivity Index), were also proposed by [3]. For both studies, the invasion areas were determined using the energy-cone model, based on the assumption of linear decay of flow energy with distance (see [85]). In both cases, the final invasion maps were only qualitative and did not account for any epistemic uncertainty quantification associated with the properties of the volcanic system and its modes and dynamics of eruption.

Where probabilities are associated with them, the above maps represent background, or long-term/base-rate, assessments of PDC hazard in the sense that none of them takes into account information and measurements that would come from monitoring and observation networks during an unrest or eruptive episode. This context is necessary because of the present difficulty of predicting the timing, size and vent location of a future eruption, based on current understanding of the state of the volcano and monitoring data. As a consequence, the definition of a quantitative background probabilistic PDC invasion map is a fundamental need not only for the aim of effective urban planning for risk mitigation, but also to have an a priori probabilistic spatial distribution of hazard to be updated during the crisis (see, for instance, [106, 148, 154]).

3.2. Methodology

Our mapping work integrates information on the distribution of the spatial probability of vent opening, the density distribution of areas previously invaded by PDCs, and the results from a simplified PDC flow invasion model. It is very difficult to predict the exact location of the next active vent as well as the scale (or typology) of the next eruptive event, so this imposes the requirement to consider the potential physical variability of these factors when producing hazard maps. Moreover, much of the available information is conditioned by large epistemic uncertainties that significantly influence the resulting maps. Also in this chapter, we implemented a doubly stochastic model able to explicitly consider, in addition to the aleatoric variability of the process, some of the main uncertainties by using structured expert elicitation techniques (see Chapters 2 and 5; *e.g.* [41, 7]). Such methods allow uncertainty distributions on identified variables to be determined for hazard assessment purposes, based on expert judgment when insufficient data or incomplete knowledge of the system do not permit conventional statistical enumeration of uncertainties.

3.2.1. Probabilistic spatial distribution of vent opening

We briefly summarize the information on the probabilistic spatial distribution of vent opening location derived from the analysis presented in the previous chapter, for making this model self-standing. In that analysis it was assumed that the probability map of new vent opening over the caldera, conditional on the occurrence of an eruption, could be expressed as a linear combination of the distribution of the eruptive vents that opened during the three epochs of recent activity of the volcano (*i.e.* the last 15 ka), the distributions of maximum fault displacement, and surface fracture density. In addition, it was considered that the probability of new intra-caldera vent opening could possibly be correlated to other variables or unidentified processes which, in this phase of our investigation, cannot be included specifically: to accommodate the influence of such unknowns, a single compensating surrogate contribution was added, assumed uniformly distributed inside the Neapolitan Yellow Tuff caldera

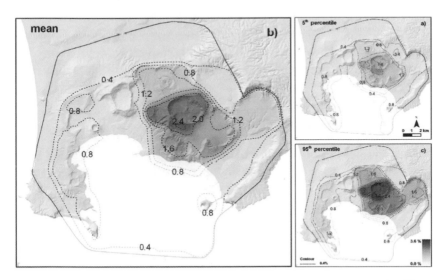

Figure 3.1. Probability maps of new vent opening location (see also Figure 2.7). Contours and colours indicate the percentage probability of vent opening per km² conditional on the occurrence of an eruption: (a), (b) and (c) refer to the 5th percentile, mean and 95th percentile values, respectively. Note that for the definition of the PDC invasion maps we do not consider the offshore portion of the caldera as a possible area of vent opening. Modified from [118].

The analysis in Chapter 2 also quantified the influence of some of the main sources of epistemic uncertainty that affect the hazard distribution. In particular, their analysis considered the uncertain localization of eruptive vents as reconstructed from field data, the number of past events which do not correspond to presently identified vents but for which stratigraphic evidence exists (*i.e.* 'lost vents'), and also the uncertain weights to be associated with variables that contribute to the definition of the vent opening map. For the definition of these weights, as well as of other relevant uncertain parameterizations, the analysis used expert judgment techniques, with a simple logic tree of target questions and various complementary procedures of structured elicitation to test the sensitivity of uncertainty quantifications to the different models.

Figure 3.1 shows an example of vent opening probability maps obtained with the above procedure (see also Figure 2.7). The three maps (Figure 3.1a, b, c) represent, respectively, the 5th percentile, the mean, and the 95th percentile of the uncertainty distribution of the vent opening location map. The distribution of past vents was computed by assuming a partitioning of the caldera into sixteen homogeneous zones, with uniform spatial distributions within each. However, very similar maps were also generated by computing a distribution of past vent positions based on Gaussian kernel functions. These synthesized results show the presence of a high probability region of vent opening in the central-eastern part of the caldera (*i.e.* the Astroni-Agnano-Solfatara area), whereas the rest of the caldera is characterized by significantly lower probability values with local secondary maxima in the Soccavo and Pisani plains and in the western zones of Averno-Monte Nuovo-Baia-Capo Miseno.

3.2.2. Distribution of PDC invasion areas

Limitations in our ability to predict the type and scale of the next eruption event require a hazard mapping approach that considers a range of possible eruptive scenarios. We take the well-known history of eruptive activity from the last 15 ka, and assume it will be representative of future patterns of behaviour of the volcano. Very large scale eruptions (*i.e.* caldera-forming), such as the NYT and the CI, are not included in the dataset due to their very low probability of occurrence (well below 1% based on frequency of occurrence estimates, *e.g.* [132, 122]).

We designated the areas invaded by PDCs as a random variable representative of the aleatoric variability affecting the next eruption event scale. The parameters of this variable were then assumed to be affected by some epistemic uncertainty as described in the following. The PDC invasion model allows us to use the area invaded by the flow as an input.

The dataset on areas inundated by PDCs that occurred in the three epochs of activity (plus Monte Nuovo) largely relies on the work of Orsi *et al.* [122], with a few minor modifications and updates due to the most recent research findings (Figure 3.2 and Table 3.1). In detail, twenty records were included for Epoch I, six for Epoch II, and twenty for Epoch III, in addition to the Monte Nuovo event (*i.e.* 47 records in total). With respect to the data of [122], the records of the events of Rione Terra and Archiaverno were deleted from Epoch I, whilst the records of Capo Miseno and Nisida were moved from Epoch I to Epoch III (see [145]). In addition, the reconstruction of the PDC distribution produced for the Agnano Monte Spina (AMS) event by [53] was considered as an alternative to the data of [122]. In all cases, part of the PDC went offshore, so on-land invasion areas were extended over the sea in order to better represent the total area affected by the flow.

The flow area boundaries (Figure 3.2) refer to the minimum areas invaded due both to the irregular distribution of outcrops and to the large erosive and anthropologic actions affecting the deposits (see [122]). Based on the available field datasets, and using alternative expert judgment procedures, the radial underestimation error of deposit boundaries (treated as epistemic uncertainty) was considered to vary between about 150 and 1,000 m, with a mean value of about 500 m (see Table 2.1). It is worth mentioning that such values are comparable with estimates of extended run-out shown by surge-like flows with respect to the underlying dense portion of the PDC, observed in recent eruptions for flows of small-medium scale (see *e.g.* [116], for more information). Such an underestimation of the deposit extent could therefore represent a minimum value of the actual flow run-out.

In order to use a representative dataset of the invasion areas, it was necessary to extend the recorded inundated areas reported in Figure 3.2 with reasonable estimates for areas of 'lost deposits'. Based on a comparison between the datasets of invasion areas (Figure 3.2) and that of the identified vents (Figures 2.2 and 2.3), a number of lost deposits were added to the three epochs. PDC invasion areas up to 10 and 50 km^2 were added as follows: for Epoch I, four records up to 10 km^2 (representing the events of Minopoli 1, Pisani 1, Fondo Riccio, Concola) and nine records up to 50 km^2 (representing the events of La Pigna 1, La Pigna 2, Gaiola, Paradiso, Paleopisani 1, S4S31, S4S32, Pignatiello 1, Casale); for Epoch II, one record up to 10 km^2 (representing the event of Baia or a flow from a lost vent); for Epoch III, seven records up to 10 km^2 (representing the events of Agnano 1, Pignatiello 2, Santa Maria delle Grazie, Paleoastroni 3, Olibano Tephra, and two PDCs from lost vents) (see [97, 145]). The choice to add lost deposits of two different areas reflects the fact that

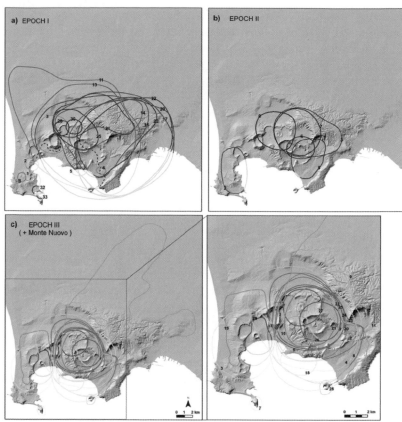

Figure 3.2. Reconstruction of the distribution of PDC deposits generated by explosive events that occurred in: (a) Epoch I, (b) Epoch II, and (c) Epoch III plus the Monte Nuovo event (modified from [118]). Numbers refer to the events reported in the legend (lines with different colour tone indicate different events). Reported deposit boundaries were extended over the sea to allow estimation of reasonable values for PDC invasion area (shown in the legend). Data were derived and updated from [122]. The distribution shown for the AMS PDC was derived from [53]. The events names and sizes are reported in Table 3.1.

the reconstruction of deposits for Epoch I is significantly more difficult than for the later epochs and so larger missing deposits are more likely to be appropriate. The spatial extents invaded by these lost PDCs were sampled using a distribution fitted to the available field datasets, truncated with the thresholds mentioned above. This information was treated as another source of epistemic uncertainty beside to the radial underestimation.

With the dataset of PDC invasion areas defined, it is possible to generate probability density functions of spatial extent distribution considering either the last 5 ka (*i.e.* Epoch III plus the Monte Nuovo event) or the last

ID	PDC Deposit - Epoch I	Area (km²)	ID	PDC Deposit - Epoch II	Area (km²)	ID	PDC Deposit - Epoch III	Area (km²)
1	Bellavista Volcano	3.9	2	Fondi di Baia Tephra	15.7	2	Agnano 2 Tephra	17.1
2	Mofete Volcano	2.1	3	Sartania 1 Tephra	40.7	3	Averno 1 Tephra	27.0
3	Gauro Volcano	16.1	5	Costa San Domenico Tephra	16.9	4	Agnano 3 Tephra	68.0
4	Santa Teresa Volcano	0.9	6	Pigna San Nicola Tephra	8.0	6	Cigliano Tephra	28.3
5	La Pietra Volcano	2.6	7	Sartania 2 Tephra	27.0	7	Capo Miseno Volcano	1.1
8	Torre Cappella Volcano	1.0	8	San Martino Tephra	19.7	8	Monte Sant'Angelo Tephra	43.6
11	Soccavo 1 Tephra	190.5				9	Paleoastroni 1 Tephra	18.1
13	Pomici Principali Tephra	129.2				10	Paleoastroni 2 Tephra	5.4
14	Paleo Pisani 2 Tephra	37.7				11	Agnano Monte Spina Tephra	312.5
16	Soccavo 2 Tephra	75.8				18	Soltatara Tephra	8.7
17	Soccavo 3 Tephra	147.5				19	Averno 2 Tephra	24.8
20	Soccavo 4 Tephra	180.2				20	Astroni 1 Tephra	39.7
21	Paleo San Martino Tephra	37.3				21	Astroni 2 Tephra	19.1
22	Minopoli 2 Tephra	113.6				22	Astroni 3 Tephra	41.1
23	Soccavo 5 Tephra	66.2				23	Astroni 4 Tephra	60.4
25	Pisani 2 Tephra	21.1				24	Astroni 5 Tephra	29.1
28	Montagna Spaccata Tephra	3.0				25	Astroni 6 Tephra	26.9
30	Pisani 3 Tephra	3.0				26	Astroni 7 Tephra	10.2
32	Bacoli Volcano	1.1				27	Fossa Lupara Tephra	8.9
33	Porto Miseno Volcano	0.7	29	Monte Nuovo Tephra	5.7	28	Nisida Tephra	4.7

Table 3.1. Names and invaded area reconstruction for PDC deposits shown in Figure 3.2. The ID values of events follow Figures 2.2 and 2.3, and their naming follows [145].

15 ka datasets (*i.e.* the three Epochs together plus Monte Nuovo). Figure 3.3a, b shows the histograms of PDC invasion areas for these two alternative datasets together with the curves of probability density functions derived from them, whilst Figure 3.3c, d shows the corresponding exceedance probability curves (survival functions). The curves, calculated by a Monte Carlo simulation, relate to the 5th and 95th percentiles and the mean coming from such uncertainty sources. In the 5 ka dataset, the AMS eruption represents an anomalous value much larger than any other record. In contrast, the presence of several intermediate data points between the body of the empirical distribution for the 15 ka dataset and the AMS event allows a quasi-continuous distribution of the PDC inundation areas to be hypothesized, and the AMS event to be considered simply one element in a continuous tail distribution, not as an extreme outlier.

The histograms (Figure 3.3) relate to the inundation areas (Figure 3.2), whilst the density and exceedance probability curves include the radial underestimation of PDC inundation areas and the randomly sampled lost deposit areas. Comparison of plots in Figure 3.3c and 3.3d indicates that the curves associated with the 5 ka history are very similar to those associated with the 15 ka counterpart, although the latter has a slightly larger number of small events and a slightly fatter tail. We assume these probability distributions valid over the whole caldera, thus neglecting any dependence of eruptive scale (*i.e.* PDC invasion area) on vent position or repose time. Following notation of Definition 3.2, we have decided to restrict $\nu_{\check{Y}}$ on the class of log-normal distributions, parameterized by the values of mean and standard deviation, assumed as random variables. Like in Definition 2.5 for assessing their distribution we rely on the uncertainty affecting the elicitation responses, through a measurable function g.

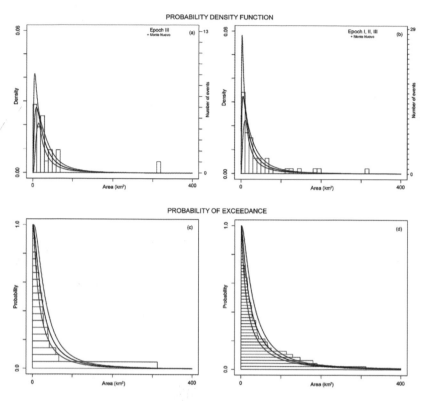

Figure 3.3. Histograms of the PDC invasion areas as estimated from Figure 3.2 for: (a) Epoch III and (b) all three Epochs (plus the Monte Nuovo event in both cases). (a) and (b) also show probability density functions for the invasion areas after consideration of underestimations of PDC run-out and the addition of 'lost deposits', as discussed in the text and Appendix A of this chapter. (c) and (d) show probability exceedance curves (survival functions) corresponding to the two periods considered, 5 ka and 15 ka. The black curve is the mean and the coloured curves are the 5th and 95th uncertainty percentiles. From [118].

Definition 3.6 (The log-normal distribution with random parameters). Let (γ_1, γ_2) be a couple of real positive random variables defined on $(E, \mathcal{E}, \eta)$; they represent the values of mean and standard deviation of a random law of PDC invaded areal sizes; the problem of assessing $\check{Y}$ is reduced to find the distribution of their block:

$$\nu_{\check{Y}}(e) = \exp_\sharp \left(\mathcal{N} \left(\gamma_1(e), \gamma_2(e) \right) \right).$$

Let $\tilde{\beta} = (\beta_j)_{j \in d'+1,...,d'+q}$ be a family of random variables defined on $(E, \mathcal{E}, \eta)$ as in Definition 2.5 representing the elicitation responses devoted to PDCs assessment. Let g be a measurable function from

$(\mathbb{R}^q, \mathcal{B}(\mathbb{R}^q))$ to $(\mathbb{R}^2, \mathcal{B}(\mathbb{R}^2))$ such that

$$g(\tilde{\beta}(e)) = \gamma(e)$$

for each $e \in E$.

The Monte Carlo simulation implemented calculates the maximum likelihood (ML) parameters for each sample, obtaining the distribution of γ and $\check{Y}$ from the distribution of $\tilde{\beta}$. The ML lognormal distribution appears to be the most defendable for characterizing available datasets; from sensitivity analyses, other different but plausible distributions do not significantly change hazard estimate outcomes (see Appendix A of this chapter).

3.2.3. Simplified PDC invasion model

The dynamics of PDCs is particularly complex due to the multi-phase nature of the flow, the highly uncertain source conditions and the complicated interactions of the current with topography (*e.g.* [59, 25, 20]). Volcanoclastic deposits at Campi Flegrei seem to be mostly characterized by surge-like facies, although a quite large variability of transport and emplacement mechanisms can be invoked for different eruptions and even for the same eruptive sequence [*e.g.* [151, 51]). Some of these complexities can be investigated by 2D/3D numerical simulations of the partial collapse of the eruption column and the propagation of PDCs over topography (*e.g.* [150, 66]). Such simulations explored, for instance, the influences of different collapsing regimes in the column and of vent positions on the PDC features. However, due to the large computation time needed to produce such simulations (of the order of some days with parallel computing), it is impractical and expensive to apply this kind of modelling within Monte Carlo algorithms involving thousands of simulations.

Therefore, with the aim of exploring main effects of the large variability of vent location and eruptive scale (*i.e.* PDC invasion area) on the area inundated, a simple integral PDC propagation model is adopted here. The model is based on the so called 'box model' of Huppert and Simpson [86] (see also [45, 80, 81]) and is suited to describing the propagation of turbulent, particle-laden currents, in which inertial effects dominate over viscous forces and particle-particle interactions. The model has been validated and calibrated through extensive comparison with 2D numerical simulations produced with the PDAC model (see [114, 64, 67]). Following notation of Definition 3.3, indeed the function F is based on such integral model, that solves a dynamical system coupling the von Kármán equation for fixed volume density currents propagation and an equation representing the particle sedimentation. A brief description of the model and its comparison with 2D numerical simulations is reported in Appendix B of this chapter; a more detailed physical description is reported in Chapter 5.

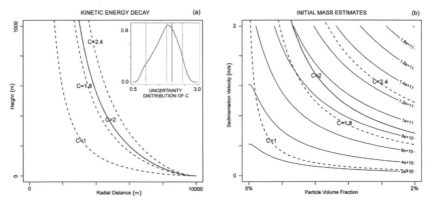

Figure 3.4. (a) Example of decay of radial flow head kinetic energy expressed in terms of potential height as a function of distance from the source. Curves refer to a flow run-out of 10 km and to values of the C parameter equal to 1.0, 1.8, 2.0, and 2.4 ($m^{2/3}$/s), as reported in the labels. In the inset the probability distribution of the C parameter is shown once a uniform distribution is assumed on the physical variables forming it. (b) Estimates of the initial mass of pyroclastic material of the column collapsing to the ground as a function of the initial concentration of pyroclasts and their sedimentation velocity. All curves refer again to a radial flow run-out of 10 km, whereas the coloured lines refer to the four values of parameter C reported in (a). See Appendix B of this chapter for more details. From [118].

The integral model allows computation of the flow kinematics and of the maximum distance (flow run-out) reached over a sub-horizontal surface by a current generated by instantaneous release (*i.e.* dam-break configuration) of a finite volume of gas and solid particles, at a given concentration. Based on outcomes of numerical simulations (*e.g.* [150, 66]), such a generation mechanism describes reasonably well the unsteady release of a portion of the column collapsing to the ground. The box model assumes that the current is vertically homogeneous and deposits particles during propagation as a function of their (constant) sedimentation velocity. No effect of wind or other atmospheric conditions is considered by the model. In the present application the model is used in its simpler formulation, which assumes a single particle size representative of the mean Sauter diameter of the grain-size distribution of the mixture. The integral model can therefore compute the flow front velocity, the average flow thickness and the particle concentration as a function of time, assuming either axisymmetric (our choice in all the examples) or unidirectional propagation, from which the kinetic energy (or dynamic pressure) of the flow front can be calculated.

In particular, in absence of topography and assuming a cylindrical symmetry with respect to the position of the vent, we calculated the kinetic energy of the flow front as a function of the distance from the vent and of the maximum run-out distance: a parameter that is equivalent to the volume of the current, and can be calculated by numerical inversion from the PDC invasion area $y \in \mathbb{R}_+$ once is fixed the vent location $x \in A$. Indeed for relying on the dataset of PDC invasion areas reported in Table 3.1, in this study the PDC invasion model is applied in an inverse mode, *i.e.* starting from the invasion area (obtained using the density functions described above) and then computing the parameter l_{max} associated with such propagation, given a specific vent location and surrounding topography (see Definition 3.7). This parameter is equivalent to the volume (or the pyroclast mass) required to generate such propagation (see also Chapter 5). However we remark that the area invaded by past PDCs was likely generated by various flows in succession, whereas our inverse analysis did not consider this aspect: additional research focused on the comparison between the box model and the 3D simulations is of the main importance.

In order to quantify main effects of topography on the propagation of a PDC, the flow kinetic energy is compared to the potential energy associated with, and therefore required to overcome, the topographical relief that the flow encounters, thus following the same approach of the energy-line (or energy-cone) model (see [85, 4]). It is worth noting, however, that the integral model allows a more realistic description of the propagation of a turbulent flow compared to the energy-line, which instead assumes a simple linear decay of flow kinetic energy more appropriate for describing the dynamics of landslides and high concentration granular flows (not commonly outcropping at Campi Flegrei).

Definition 3.7 (The kinetic energy inequality). Let K be a real positive measurable function defined on $[0, \text{diam}(B)] \times \mathbb{R}_+$, such that $K(r, l_{max})$ represents the kinetic energy at a distance r from the vent, neglecting topography for the flow front of a PDC with maximum run-out l_{max}. Let $\tilde{F}$ be a functional from $A \times \mathbb{R}_+$ to the Borel subsets of B such that

$$\tilde{F}(x, l_{max}) := \{z \in B : K(d(z, x), l_{max}) - U(z) > 0\}$$

where d is the Euclidean distance on $B \subseteq \mathbb{R}^2$, U is a real measurable function defined on $(B, \mathcal{B}(B))$ and representing the potential energy associated with the local topography. Such a set $\tilde{F}(x, l_{max})$ in a first approximation represents the set of points that the current has enough kinetic energy to reach.

Figure 3.4a illustrates the non-linear decay of the flow kinetic energy, expressed as potential height for a radial flow with run-out 10 km, as a function of distance from source and as a function of the C parameter $(m^{2/3}/s)$ which accounts for the initial volume concentration of particles and their sedimentation velocity in the flow (see Appendix B of this chapter). The main approximation is to compare the potential energy associated to the topographical barriers in position z to the kinetic energy calculated neglecting the topography that separates z and the vent x. We modify the definition of $\tilde{F}$ to avoid at least the situation of jumps in the propagation. An additional important aspect is that the PDC invasion model is applied in an inverse mode.

Definition 3.8 (The inversion of invaded area). Let σ be a set function from $A \times \mathcal{P}(B)$ on $\mathcal{P}(B)$, such that for each $D \subseteq B$, $\sigma(x, D)$ is the subset of D that is a star-convex set with respect to x. Let R be a real positive measurable function defined on $A \times \mathbb{R}_+$, such that the value $R(x, y) = l_{\max}$ represents the maximum run-out as a function of the vent location $x \in A$ and the PDC invasion area $y \in \mathbb{R}_+$. We finally define an expression for the functional F

$$F(x, y) := \sigma(\tilde{F}(x, R(x, y))).$$

We remark that it was possible to calculate the values of the function $R(x, y) = l_{\max}$ by numerical inversion of the expression

$$\int_B 1_{\sigma(\tilde{F}(x, l_{\max}))}(z)dz = y$$

that is in principle a discontinuous, but monotone crescent in function of $l_{\max}$. More details can be found in Appendix C of this chapter. Figure 3.4b shows the mass of pyroclasts collapsing to the ground that is able to generate a radial PDC with run-out 10 km, as a function of the initial pyroclast volume concentration and sedimentation velocity, considering a flat topography. The physical parameters of the flow adopted in the integral model are assumed representative of the eruptive mixture and collapse conditions at Campi Flegrei (*e.g.* [51, 150, 65, 66]).

3.3. Results

Combining the spatial probability map of new vent opening, the probability distribution of PDC invasion areas, and the PDC integral box model described above, it is possible to produce several probabilistic hazard maps of PDC invasion with their associated uncertainty. In the following, a few cases are shown and discussed to illustrate our main findings.

Several other maps were produced to investigate the influence of some key variables or assumptions on the hazard mapping. Our maps relate solely to the probability of invasion by PDCs and not to the distributions of specific hazard variables, such as dynamic pressure and temperature. We also assumed that a future PDC episode will originate in the on-land portion of the caldera because source conditions would be fundamentally and significantly different in the case of an underwater vent.

Our invasion maps are the result of a Monte Carlo simulation procedure, implemented to combine the several probability distributions discussed above (aleatoric variabilities) together with their epistemic uncertainties, based on a doubly stochastic model. The Monte Carlo simulation has a nested structure, configured for estimating uncertainty on the results: as a consequence, the procedure creates maps of PDC invasion in terms of a mean (or median) value and of representative percentiles with respect to the uncertainty sources we consider. With the location of the eruptive vent determined and the areal size to be invaded by the flow defined, the simulation of a single PDC propagation event associates a value of 1 to those zones reached by the flow, and 0 otherwise. This is done using the PDC flow model in inverse mode and including the blocking effect of the topography (see Appendix C of this chapter). Therefore, for each outcome of the epistemic uncertainty sources (*i.e.* uncertainty on the probability map of new vent opening, uncertainty on the density distribution of the PDC invasion area), by repeating the simulation of a single PDC a large number of times randomly changing vent location and inundation area, and then aggregating the zone 0/1 values obtained to estimate their means, it is possible to approximate, by the law of large numbers, the probability that each location of the map is reached by a PDC conditional on the occurrence of an explosive eruption. To limit computation time, most of the maps are produced using a regular Cartesian grid of cells with 500 m sides, although simulations were also performed with higher resolutions to investigate the sensitivity of results to this numerical parameter. For instance, a grid of cells with 250 m sides produced maps similar to those using 500 m resolution. Based on the 500 m grid, and due to the nested structure of the Monte Carlo procedure, each invasion map requires the execution of over half million model simulations.

Figure 3.5 shows the PDC invasion probabilities in terms of a mean map and maps of the 5th and 95th percentiles, assuming the vent opening probability maps of Figure 3.1, and the probability distribution of invasion areas associated with the 5 ka dataset (See Figure 3.3a, c). The PDC invasion probability maps assume that each new event would be able to produce PDCs just from a single vent located in the on-land portion of the caldera. With reference to the mean map, from the distribution of isolines

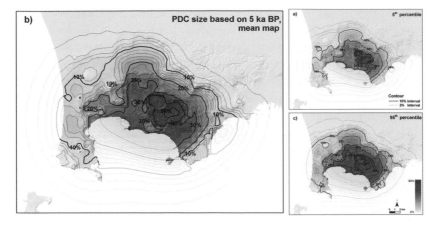

Figure 3.5. PDC invasion probability maps computed by assuming the vent opening distribution described in Figure 3.1 and the density distribution of invasion areas of the last 5 ka, shown in Figure 3.3a. The maps assume that PDCs originate from a single vent per eruption, and that the vent is located in the on-land part of the caldera. Contours and colours indicate the percentage probability of PDC invasion conditional on the occurrence of an explosive eruption. The maps relate to: (b) the mean spatial probability, and (a) the 5th and (c) 95th percentiles, respectively. Modified from [118].

of equal invasion probability it emerges that, consistent with the deposit data (*e.g.* [122]), the central-eastern part of the caldera is the most exposed to PDC hazard with peaks of probability of invasion of about 53% in the Agnano plain. Probabilities above 45% are also computed in the Astroni area and above 30% in the area of Solfatara. Note that mean probability values above 5% apply to almost all parts of the caldera (with the exception of a small portion of the Capo Miseno peninsula) and a large part is associated with mean values exceeding 10% PDC invasion probability. Values between about 5 and 10% affect some areas outside the caldera border (*e.g.* Collina di Posillipo and some neighborhoods of Naples). The plots showing 5th and 95th percentiles (Figure 3.5a, c) enumerate the substantial uncertainty on these mapped probabilities of PDC invasion with respect to the sources of epistemic uncertainty described above. These maps apply only to the propagation of PDCs over the landward portion of the caldera; isolines of the probability of invasion in offshore parts of the caldera are shown in outline to give a first approximation of the potential hazard represented by PDCs traveling over the sea. We assume the sea surface as flat ground topography with no effect of the water on the PDC propagation (thus neglecting any specific heat and mass transfers between flow and sea), although theoretical studies have pointed to a reduced mobility of PDCs over water (*e.g.* [60]).

Figure 3.6 shows the same maps as Figure 3.5 but in this case assumes the probability distribution of invasion areas derived from events over the last 15 ka (see Figure 3.3b, d). The maps are very similar to those of Figure 3.5, with a slight increase in areas affected by low probabilities of invasion (see, for instance, the 2-10% isolines) and an associated slight decrease of the peak values computed in the central-eastern part of the caldera. As mentioned above, these effects are the result of the slightly fatter tail of the distribution related to the 15 ka history compared to that of the 5 ka record (see Figure 3.3).

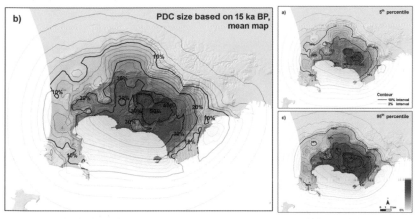

Figure 3.6. PDC invasion probability maps computed by assuming the vent opening distribution described in Figure 3.1 and the density distribution of invasion areas of the last 15 ka, shown in Figure 3.3b. The maps assume that PDCs originate from a single vent per eruption, and that the vent is located in the on-land part of the caldera. Contours and colours indicate the percentage probability of PDC invasion conditional on the occurrence of an explosive eruption. The maps relate to: (b) the mean spatial probability, and to (a) the 5th and (c) 95th percentiles, respectively. Note that the colour scale used in these maps is consistent with that used in Figure 3.5. Modified from [118].

Several other maps, for the sake of brevity not described here, were also produced, some of which are reported in Figure 3.7. These investigate the effects on the PDC invasion probability maps of: i) the procedure followed for the definition of the vent opening map (*i.e.* kernel function *vs* partitioning of the caldera, see Chapter 2, for explanations) (Figure 3.7b); ii) neglecting fault-, fracture- and homogeneous distribution maps in the definition of the vent opening map (Figure 3.7c); iii) considering only the vent locations of Epoch III in the definition of the vent opening probability map (Figure3.7d); iv) different physical properties of the flows adopted in the PDC invasion model (Figure 3.7e), and v) the vent opening probability map of Selva *et al.* [141] (Figure 3.7e). Some more

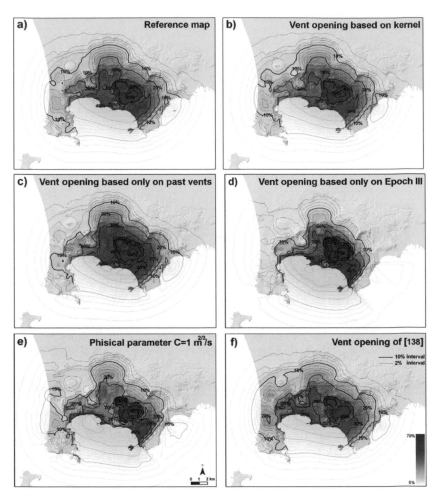

Figure 3.7. Ensemble of mean spatial probability maps of PDC invasion showing the effect of different assumptions. (a) Reference mean map (as in Figure 3.5b) assuming the vent opening map of Figure 3.1 and the density distribution of PDC invasion areas of the last 5 ka in Figure 3.3a; (b) PDC invasion mean probability map assuming the vent opening maps of Figure 2.7a, b and c, based on the use of kernel functions; (c) mean map obtained by neglecting the influence of faults, fractures and homogeneous distribution maps on vent opening probability; (d) mean map obtained considering only the distribution of vent location of the events of Epoch III; (e) mean map obtained by assuming a value of parameter C equal to 1 $m^{2/3}$/s instead of 2 $m^{2/3}$/s , as assumed in all other maps (see Appendix B of this chapter), and (f) mean map obtained by assuming the vent opening map of [141]. Contours and colours indicate the percentage probability of PDC invasion conditional on the occurrence of an explosive eruption from a single vent located on-land. Note that the colour scale is consistent with those used in Figures 3.5 and 3.6. Modified from [118].

details on the specific parameters used to produce these maps are given in the caption to the figure. As it emerges from the comparisons, despite some significant differences observed locally in specific areas of the caldera, our main findings about the spatial distribution of PDC invasion probabilities remain largely valid.

Our method also enables us to draw probabilistic invasion maps that consider eruption events constrained below a defined upper scale limit (in our analysis, up to a defined PDC invasion area). This can be obtained straightforwardly by truncating the eruptive scale distributions of Figure 3.3c, d at a given limit. For instance, Figure 3.8 shows PDC invasion maps representative of the mean, and 5th and 95th percentiles of the distribution, when the limiting value corresponds to 5% exceedance probability of the PDC invasion areas with reference to the 5 ka curve (corresponding to 112 km^2 areal size, $\pm$15 km^2; Figure 3.3c). Based on the estimates of the probability of occurrence at different eruptive scales, as computed by [123], such a limit approximately corresponds to the occurrence of explosive eruptions of small and medium scale, but not large scale events (*e.g.* the AMS eruption; the average probability of occurrence of large eruptions is, in fact, estimated to be about only 4% of all scale sizes). Under this restriction, the resulting PDC invasion maps (Figure 3.8) remain similar to those that consider the full distribution of eruptive scales (Figures 3.5 and 3.6). However, in Figure 3.8 the probability isolines now affect slightly smaller areas due to the neglecting of PDCs produced by large scale events. Of course, maps of the same type could be produced for other thresholds associated with other probabilities of exceedance of the PDC invasion areas (Figure 3.3c, d).

Finally, we investigate the possibility of simultaneous or near-simultaneous openings of multiple vents in zones of the caldera significantly distant to each other (*i.e.* not related to vent migration within the same area). Such occurrences are indeed a possible scenario at a caldera, as shown at Rabaul volcano (Papua New Guinea) in September 1994, with the simultaneous opening of the vents of Tavurvur and Vulcan volcanoes, on opposite sides of the caldera (8 km distant from each other) (see [130]). Recent work (*e.g.* [97]) has shown that such a phenomenon likely occurred also at Campi Flegrei with the contemporaneous eruption, about 4.3 ka BP, of the Solfatara and Averno centers, located about 5.4 km apart (*i.e.* the events of Solfatara and Averno2). Tephra placed at the same height of the stratigraphic record (*e.g.* [57, 145]) and still not chemically and physically correlated suggest that other groups of eruptions could have been simultaneous. Multiple venting implies therefore the possibility of an increased area potentially invaded by PDCs in an eruptive episode, both inside and outside the caldera.

Based on all the available evidences at Campi Flegrei and elsewhere, the probability of the opening of two simultaneous vents is estimated from expert judgment to be about 10%, but with an uncertainty range from about 5% to 25% (corresponding to the 5th and 95th credible range percentiles, see Table 2.1 for details). Based on these numbers, Figure 3.9 shows PDC invasion probability maps for the scenario of two simultaneous vents in terms of (b) the mean map, and (a) 5th and (c) 95th percentiles. No constraint was imposed on the distance between the two simultaneous vents. Based on the probability map of vent opening, a mean distance between dual vents of 4.7 km was calculated (assuming two independent samples from the same spatial distribution), with 5th and 95th percentiles of 1.0 and 10.0 km, respectively.

The maps of Figure 3.9 are comparable to those of Figure 3.5 in the sense that they assume, for both eruptive centers, the same probability of vent opening of Figure 3.1 and the probability density function of the PDC invasion areas of the last 5 ka. In this scenario, the area invaded by the flows generated by two simultaneous vents is computed as the union of the areas invaded by the PDCs that originate from the two distinct vents. From comparison with the maps of Figure 3.5, it emerges that in the scenario with dual venting the peak probabilities computed in the Agnano plain are about 5% higher and also that the isolines representative of the 5% and 10% probability of invasion occupy, in this situation, slightly wider areas.

3.4. Discussion

In this study, we develop an innovative method to generate probabilistic maps of PDC invasion in caldera settings conditional on the occurrence of an explosive eruption. Our approach allows different strands of data to be combined within a probabilistic framework and, most importantly, enables us to consider and quantify the influence of some key sources of epistemic uncertainty present in the volcanic system. The approach is particularly relevant for caldera settings due to the large variations of possible vent locations and eruption scales that can be exhibited by volcanoes of this type (aleatoric variabilities).

In the present case of Campi Flegrei, PDC invasion maps are obtained by conflating a probabilistic distribution for new vent opening position, a distribution of PDC invasion areas assumed representative of the range of eruption scales (based on an updated version of the dataset of [122]), and a simplified PDC invasion flow model able to account for the PDC scaling properties and the main effect of caldera topography on the extent of areas invaded by the flows. These probabilistic distributions are also able to account for some of the main epistemic uncertainties affecting the volcanic system (see also Chapter 2). In particular, the analysis takes account of the uncertain location of past vents, the number of 'lost vents', the uncertain correlations between the distribution of observable features of the caldera, such as faults and fractures, and the spatial probability of vent opening, the incomplete reconstruction of areas invaded by previous PDCs, and the possibility to have simultaneous activation of two distinct vents during the same eruptive episode. Our analysis relies on evidence about the last 15 ka of activity of the volcano and therefore does not include extreme caldera-forming events such as the CI or the NYT eruptions. Moreover, all the maps presented here presume that the eruptive vent openings take place in the landward portion of the caldera; offshore eruptions are, fundamentally, a different, and more difficult, problem to tackle.

We provide PDC invasion maps under different assumptions in order to investigate their relative relevance and the robustness of the results. Assuming the activation of a single vent per eruptive event, it emerges from these maps that the whole caldera is significantly exposed to PDC hazard (*e.g.* Figure 3.5). Mean invasion probabilities above 5% are calculated over almost the whole caldera, with peak values just exceeding 50% in the Agnano plain. The areas of Astroni and Solfatara are exposed with mean values above about 30%. Mean probabilities of about 10% are also computed in some areas outside the caldera, in particular over Collina di Posillipo and in some neighborhoods of the city of Naples. Consideration of the density distribution of PDC invasion areas over the last 15 ka (see Figure 3.6) does not affect significantly the probability distribution described above but just extends slightly the area affected by low probability isolines, simultaneously reducing slightly peak value probabilities in the central-eastern part of the caldera. Different assumptions about the vent opening mapping and PDC properties also produce changes to the probability values of about the same amount, as shown in the additional maps reported in Figure 3.7.

These maps also allow the influence of different eruption scenarios to be considered. Figure 3.8, for instance, relates to the possibility to define an upper limit on the expected eruptive scale of a future event. Specifically, the probability distribution of the PDC invasion areas (Figure 3.3) was restricted to its 95th percentile value to produce Figure 3.8. This limit represents approximately the occurrence of small to medium scale events at Campi Flegrei, but not large scale events (such as the AMS event, see [123]). Under this constraint, the computed distribution of probability results is again very similar to that described above, but in this case with a general decrease in mean values of about 2%. Nevertheless, essentially the whole caldera is still characterized by mean probabilities of flow invasion larger than 5%, and values up to about 10% are again computed in some eastern areas outside the caldera rim.

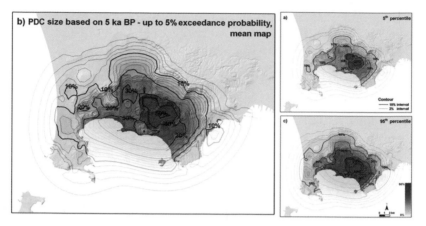

Figure 3.8. PDC invasion probability maps computed by assuming the vent opening distribution described in Figure 3.1 and the density distribution of invasion areas of the last 5 ka shown in Figure 3.3a with a bounding limit corresponding to 5% exceedance probability for invasion area, *i.e.* 112 ± 15 km^2. The maps assume that PDCs originate from a single vent per eruption, and that the vent is located in the on-land part of the caldera. Contours and colours indicate the percentage probability of PDC invasion conditional on the occurrence of an explosive eruption. The maps relate to: (b) the mean spatial probability, and to (a) the 5th and (c) 95th percentiles, respectively. Note that the colour scale used in these maps is consistent with those used in Figures 3.5, 3.6 and 3.7. Modified from [118].

Similarly, Figure 3.9 considers the possibility of simultaneous activation of two separate vents during the same eruptive event. This possibility has been postulated as having happened already at Campi Flegrei (see [97]), and has the effect of increasing the area potentially affected by PDC invasion. Assuming this scenario could occur in 10% of all eruption episodes, with a credible range between about 5% and 25%, the resulting mean invasion map produces slightly wider inundation footprints with a general increase of probability values of about +2% compared to the case of single vent.

An important outcome of our approach is the possibility to identify and quantify some of the sources of epistemic uncertainty affecting the phenomena of concern. This permits us to generate not only a mean (or expected value) map of the probability of PDC invasion but also a set of maps that represent 5th and 95th percentile uncertainty spreads. From inspection of the results, the difference in relative percentage between the 5th or 95th percentiles and the local mean values (*i.e.* divided by such

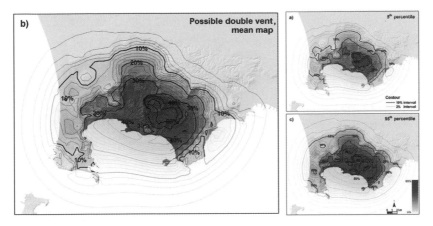

Figure 3.9. PDC invasion probability maps computed for PDCs that possibly originate from two simultaneous vents in an eruptive event, with the vents located in the on-land part of the caldera. The calculations assume the vent opening distribution described in Figure 3.1 and the density distribution of invasion areas of the last 5 ka, shown in Figure 3.3a. The probability of a double event is 5% - 15% - 25%. Contours and colours indicate the percentage probability of PDC invasion conditional on the occurrence of an explosive eruption. The maps relate to: (b) the mean spatial probability, and to (a) the 5th and (c) 95th percentiles, respectively. Note that the colour scale used in these maps is consistent with those used in Figures 3.5, 3.6, 3.7 and 3.8. Modified from [118].

mean values) can be approximately quantified inside the caldera typically as ±25% of the mean probability values, with variability from about ±15% up to ±35% (corresponding to the 5th and 95th percentiles) in different areas of the caldera. Outside the caldera the average variability rises to about ±55% of the local mean, with ranges from about ±30% to ±110% from place to place. Despite the significant sizes of such uncertainty estimates, in the present analysis just some of the relevant sources of epistemic uncertainty were considered, as previously described. Other possible influences, for instance, dependence of vent location and temporal patterns on eruptive scale, the effect of eruption duration, the accuracy of the PDC propagation model, complexities of 3D topography on flow propagation, as well as the potential influence of atmospheric conditions, are not included in the present analysis and could represent objectives of future studies.

The limitations of the PDC propagation model and of the stopping criterion should be considered when evaluating the invasion maps. The integral box model does not take into account complex processes occurring during PDC propagation, such as partial blocking of the current by topographical barriers, the generation of buoyant thermals and co-ignimbrite columns from disruption of the main flow or by passing over topographic obstacles, and the complex multi-dimensional and transient effects associated to the interaction of a flow with the ground topography (see [150, 66]). The effect of wind on the propagation of the PDC is also neglected. Moreover, our maps are computed on a Cartesian grid with cells of side 500 m, meaning that the associated probability should be interpreted as a mean value over the cell space and that details below this scale are not meaningful. This is the case, for instance, for some small island-shaped probability contour areas located mostly over Collina di Posillipo and eastwards that are generated by the complex interplay between the envelope of all simulations with varying vent location and scale and the rough topography of the caldera. As a consequence, detailed local-scale zonation of the flow invasion probabilities cannot be achieved using the approach illustrated here. For that purpose, more accurate transient and multi-dimensional physical models and more detailed analyses of local topography should be used.

The probabilities of flow invasion reported in all the maps (Figures 3.5 - 3.9) are conditional, as mentioned above, on the occurrence of an explosive eruption from a vent or vents in the subaerial portion of the caldera. This means that to compute the probability of invasion conditional on the occurrence of an unspecified eruption (*i.e.* effusive or explosive, with vents located on-land or offshore) it is necessary to multiply all the probability isoline values by (1-P), where P is the probability of being effusive (assuming, for the sake of simplicity, an equal vent opening spatial distribution for explosive and effusive eruptions) or the probability of having a vent located in the sea (hence not producing a significant PDC hazard in the common sense). By assuming a probability of occurrence of an effusive eruption of about 10% (see [123]), and a probability of occurrence of an eruption with vent located offshore of about 25% (see Chapter 2), the probability values reported on the maps presented here need to be multiplied by a factor (1-P) of about 0.68 (assuming the two circumstances to be independent). Finally, it is possible to highlight the notable probability that a Campi Flegrei PDC originating on-land would likely interact with seawater. Significantly wide areas along the coast of the municipality of Pozzuoli have associated mean probabilities of flow invasion up to about 40%, with all the coast of the Golfo di Pozzuoli being potentially affected with mean probabilities above 10%. The generation of a PDC-induced tsunami should therefore be considered a possibility, such as was observed during the 1994 eruption of Rabaul (see [119]) and the eruptive crises of the Soufrière Hills volcano, Montserrat (*e.g.* [109]). This adds the hazards associated with PDC-induced tsunami waves to those of other hazardous processes generated by potential explosive events with a vent located offshore, which possibility is estimated to have a mean probability of occurrence of about 25% (based on Chapter 2 results).

3.5. Appendix A: Classes of distributions representing the PDC invasion areas

In order to choose which distribution fits better the datasets reported in Figure 3.3, *i.e.* the 5 ka dataset (D1, Figure 3.3a, c) or the 15 ka dataset (D2, Figure 3.3b, d), we performed some analyses and statistical tests. In particular we focused on the maximum likelihood (ML) lognormal, the ML Weibull and the Pareto distributions.

Recall that the density function of a lognormal distribution of log-mean m and log-standard deviation s, is:

$$f_{lN}(x) = \frac{1}{x\sqrt{2\pi s^2}} \exp\left(-\frac{\log^2(x-m)}{2s^2}\right) = \frac{s\sqrt{2}}{x\sqrt{\pi}}(x-m)^{-\log(x-m)}.$$

whereas the density function of a Weibull distribution of mean $\lambda > 0$ and shape $k > 0$ is:

$$f_W(x) = \frac{k}{\lambda^k} x^{k-1} \exp\left(-(x/\lambda)^k\right)$$

In the first case, the logarithm and the exponential terms of the expression counterbalance each other to a certain extent and produce a quasi-polynomial decay although faster in the limit. In the second case, the distribution produces a quasi-exponential tail although slower in the limit. While both distributions fit quite well the body of the datasets, the lognormal distribution gives higher likelihood to the largest values. Statistical analyses were performed in order to quantitatively evaluate the effect of this choice.

Some criteria were unable to discriminate between the two distributions. The Akaike information criterion values (*i.e.* the logarithm of the maximal likelihood) are very similar for both datasets and therefore could not provide discerning indication. Similarly, a measure of fitting error was unable to give a clear difference between the two distributions. We define the fitting error E as the L^1 distance between the cumulative function of our estimation choice and the cumulative empirical function of the observed data, rescaled in proportion of the range of the dataset. In the case of dataset D1, the fitting errors of the ML lognormal and ML Weibull resulted in values about 4.8×10^{-2} and 5.5×10^{-2} respectively; conversely for the dataset D2, values of distance of about 4.6×10^{-2} and 2.6×10^{-2} were, respectively, obtained for the two distributions.

Therefore, in order to find the best distribution to use, a statistical test that estimates how probable the observed values are, supposing they are extracted from an ML lognormal distribution or from an ML Weibull distribution, was carried out. By using simple Monte Carlo simulation, the distribution of the index E when the observed datasets are substituted with a random sample of the same size extracted from each of the ML distributions was determined. The calculated p-value is the probability of extracting a statistical sample that produces a fitting error E greater than that associated with the actual data: therefore, a very small p-value means that is improbable to find the real dataset with that distribution and, in contrast, a large p-value means that the distribution is a good candidate to generate realization values similar to those observed. With dataset D1, p-values of about 0.45 and 0.17 were obtained for the ML lognormal and ML Weibull, respectively, whereas p-values of 0.94 and 0.7 were obtained for the D2 dataset, respectively. Based on this test, the ML lognormal is therefore preferred to the Weibull distribution, and seems to fit better the larger elements of the body of the data.

The fact that the ML lognormal distribution fits the tail of the distribution better than a ML Weibull suggests that the tail behaviour is more nearly polynomial rather than nearly exponential. The representative class of probability measures that have density functions with polynomial tails is the Pareto (power laws), with typical density expression as:

$$f_P = \frac{\alpha x_0^\alpha}{x^{\alpha+1}},$$

for all $x > x_0$, and 0 otherwise, the two parameters representing the exponent $\alpha > 0$ and the threshold $x_0 > 0$. In order to test this type of distribution, the datasets were separated also into two subsets to estimate separately the body and the tail of the distributions. Adjusting the choice of x_0, a joint Weibull-Pareto distribution was fitted to the data. However, due to the small number of data that define the behaviour of the tail, even in the case of the full dataset D2, this approach was not able to provide better results with respect to the ML lognormal distribution. Due to the above considerations the ML lognormal was assumed in our analysis.

3.6. Appendix B: Pyroclastic density current box model

The box model of Huppert and Simpson [86] allows the kinematic properties of a PDC to be computed under the assumption that a given volume of pyroclastic mixture is instantaneously released and the flow is assumed vertically homogeneous (*i.e.* turbulent and well-mixed) and traveling on a sub-horizontal surface. These assumptions allow a simple dynamical system to be stated, providing a relationship for the rate of propagation, depth and average particle concentration of the current as a function of time (see Chapter 5 for more details). If $u(t)$ is the velocity of the front of the current, $l(t)$ is its position and an axisymmetric propagation of the flow is assumed, the model states:

$$\begin{cases} u = \frac{dl}{dt} = Fr \left(g_p \phi h \right)^{1/2} \\ \frac{d\phi}{dt} = -w_s \frac{\phi}{h}, \\ l^2 h = V. \end{cases}$$

where Fr is the Froude number, g_p the reduced gravity, ϕ the volume fraction of particles in the flow, w_s the sedimentation velocity of particles and V the volume of collapsing mixture divided by π. After simple computations we find the function

$$l(t) = [\tanh(t/\tau))]^{1/2} l_{\max}$$

solves the above equation, where $\tau = \left(Fr^{-1} (g_p \phi_0 V)^{-1/2} l_{\max}^2 \right)/2$, $\phi_0 = \phi(0)$ is the initial volume concentration of particles in the mixture, and $l_{\max}$ is the maximum distance reached by the flow (*i.e.* the PDC run-out) that it is possible to calculate from the other parameters as:

$$l_{\max} = \left(8\phi_0^{1/2} g_p^{1/2} V^{3/2} w_s^{-1} Fr \right)^{1/4}.$$

In a similar way to the energy-cone approach, the front average kinetic energy is computed and compared to the potential energy associated to an obstacle of height H. Here the comparison was done considering simple gravity and also neglecting hydraulic effects associated with flow-obstacle interactions. By using the above formula of $l(t)$, we derive an expression for $u(l)$ and therefore the following function for H:

$$H = \frac{1}{2g} \left[\frac{C l_{\max}^{1/3}}{x \cosh^2 \text{artanh}(x^2)} \right]^2, \quad C = \left(Fr^2 w_s \phi_0 g_p \right)^{1/3} /2.$$

where $x = l/l_{\max}$. This function basically replaces the straight line of the energy-cone model. It should be noted that the parameter C is the only physical parameter of the integrated model and therefore its value can be obtained with different combinations of the variables that form it. Assuming reasonable bounds on the physical parameters involved for the Campi Flegrei case, such as $w_s = 0.05 - 1.2$ m/s (corresponding to mean Sauter particle sizes between 10 and 500 μm), $Fr = 1 - 1.19$ (as resulted from calibration tests), $\phi_0 = 0.5 - 1.5\%$, and $\rho_p = 700 -$

1000 kg/m^3, and assuming a uniform probability distribution between these ranges as appropriate given the large uncertainty affecting these parameters, the mean and median values of the C parameter result around $2 \text{ m}^{2/3}/\text{s}$ (1.0, 1.8 and $2.4 \text{ m}^{2/3}/\text{s}$ corresponding to the 5th, 50th and 95th uncertainty percentiles respectively; see the inset of Figure 3.4a for the C uncertainty distribution). Therefore, in most of the simulations a value of $2 \text{ m}^{2/3}/\text{s}$ was assumed although, to check the effect of this variable on the spatial probability map, a value of $1 \text{ m}^{2/3}/\text{s}$ (corresponding to the 5th percentile) was also used, as shown in Figure 3.7e. Adopting instead the other extreme value of $2.4 \text{ m}^{2/3}/\text{s}$ (corresponding to the 95th percentile) does not produce significant changes to the corresponding mean map. Figure 3.4 shows that a value of $C = 1 \text{ m}^{2/3}/\text{s}$ is representative of a PDC, at constant any other initial variable, richer of fine particles and therefore more mobile (*i.e.* able to reach a specific run-out distance with a lower amount of collapsing mass) than PDCs with a value of $C = 2 \text{ m}^{2/3}/\text{s}$ or greater.

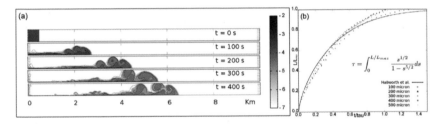

Figure 3.10. (a) Isocontours of the Logarithm base 10 of particle volumetric fraction of the current with particles of diameter $dp = 100\mu m$ and initial particle volume fraction $\phi_0 = 5 \times 10^{-4}$; (b) non-dimensional front position versus non-dimensional time for particle-laden currents with different values of the Sauter particle diameter. For each particle class the dimensionless scaling law was obtained setting the Froude number $Fr = 1.18$ and computing the settling velocity from the free particle fall in a still current with equal volume fraction. Modified from [118].

The integral box model has been extensively tested against laboratory experiments (*e.g.* [72]) and numerical simulations able to describe the dynamics of stratified PDCs. In particular, the model was validated in non-ideal conditions, *i.e.* in case of significant density differences between the flow and the ambient and assuming different particles sedimentation rates. As an example, Figure 3.10a shows the time evolution of the flow calculated by the numerical model PDAC (see [64, 66]) with density contrast of 0.4 and particles of 100 μm diameter, whereas Figure 3.10b shows a comparison between the numerical model results and the box model predictions in case of sedimenting currents with different particles sizes (from 100 to 500 μm) in Cartesian coordinates. When set to non-dimensional variables (*i.e.* t/τ and l/l_{max}) all simulations collapse on to the curve predicted by the box model, confirming minor influences on flow propagation by current stratification and viscous and buoyancy forces.

As mentioned in the main text, the model is applied in an inverse mode in order to produce the invasion maps (see also Appendix C of this chapter). This means that the model is used to estimate the mass (or the equivalent volume) of the collapsed pyroclastic mixture able to invade the inundation area, as extracted from the density functions derived from field reconstructions. Given a specific vent location and associated surrounding topography, such a calculation is carried out numerically by an iterative procedure based on the secant method, with an initial condition estimated from inversion of a simple energy-line model. The method reproduces invasion areas with a relative error below 0.05 in 95% of cases and with just about 5 - 6 iterations. Calculation of the area invaded by the PDCs is also computed adopting different grid resolutions and numerical algorithms. For instance, the invasion areas of a single PDC can be obtained assuming both a regular Cartesian grid up to 50 m resolution and a radial discretization of the space in 360 sectors by using a 10 m Digital Elevation Model resolution. Different assumptions were also made on the way topographic reliefs shade the downstream areas with specific reference to the algorithms implemented to compute the areas; results indicate some effects of these choices on some limited areas of the final hazard maps that, however, can be quantified to the order of few percentage points in terms of probability of flow invasion.

3.7. Appendix C: Details of the implemented propagation algorithm

In the sequel are presented more details about the simplified procedure that is implemented to simulate a single PDC propagation: it is repeated thousands of times during a Monte Carlo simulation. This and similar models, with the main purpose of being very fast at the cost of some accuracy, are also called emulators. An additional aspect is the assumption of the areal size to invade from the beginning, instead of more usual initial conditions about volume (or mass) of the flow, although there are simplified relations linking the two parameters. For this reason our propagation model is structured in two parts: the (direct) simulation of a flow of fixed initial conditions and then a numerical inversion aimed at obtaining a simulation that reaches a pre-imposed areal size. A third paragraph is devoted to the implementation of the Monte Carlo simulation.

Direct simulation A trivial way to cope with this problem is not implementing the topography into the algorithm: just assuming circular areas of invasion for the flows gives results that a posteriori are not so far from the ones from relatively more accurate other simple models of invasion. The motivation of such robustness is the adoption of the invaded area sizes as the main input: the algorithm seeks for the shape of the area, not its size. Merely linking the areal sizes to the areas of past deposits permits to obtain a model that is indirectly influenced by (past) topographical barriers. Another traditional approach that is commonly adopted to take into account the influence of the topography is the well known energy line model (see Hsu (1975)). The core of the energy line model is the following idea: suppose to set a bidimensional domain B, and a point $(x_1, x_2) \in B$ that represents the coordinates of the eruptive vent; once it is fixed the height H of the eruptive column, its collapse is represented on each radial line on B through the profile K of kinetic energy of the front of the flow, defined as a linear function of the distance $d(i, j)$ of each point (i, j) from the vent. The only other physical parameter is the slope coefficient α of such line. This approximation is unidimensional, but the kinetic energy profile forms a cone in space: for this reason it is also called energy cone model. This approach gives better results for simulating granular flows, as in the case of landslides and avalanches.

The topography effect is implemented through a Digital Elevation Map (DEM), *i.e.* a matrix containing the elevation of each point (i, j) above sea level. The function

$$\mathcal{H}(i, j) = K(i, j) - g \cdot DEM(i, j),$$

where the mass of the flow front is simplified, and simple gravity g is considered, compares the kinetic energy with the potential energy needed for the overcome of topographical barriers. This permits to discriminate the points in two classes: when $\mathcal{H}(i, j) > 0$ the flow in principle has enough energy to reach and invade the point (i, j), otherwise (i, j) cannot be reached because the flow does not possess enough energy. Moreover, for better representing such stopping on each single radial line, the model does not invade any point farther than the first point with negative $\mathcal{H}$: in other words the model does not invade the points shaded by any topographical barrier respect to the vent.

The model is summarized in these two phases: the energy comparison and the shading of topography barriers; in the particular topography of a caldera the second phase becomes particularly important for deciding in what way the flow interacts and possibly overcomes the caldera boundary. The implementation of the box model is build through the same two phases, but in the energy comparison phase our approach changes the radial profile of the kinetic energy: instead of using a straight line it is adopted the nonlinear curve of decay obtained from the box model physical approximation (see Figure 3.4a, Appendix B of this chapter, and also Chapter 5). More alternative procedures have been explored for optimally solving the shading problem, and their implementations are reported in Chapter 6. The first approach is dividing the space in $N = 360$ circular sectors with center on the vent and then assuming invaded the totality of each sector from the vent up to the distance of the first negative value of $\mathcal{H}$ restricted on its bisecting line. This approach is the traditional one for the energy line models. Another shading procedure was also implemented, setting a discrete grid and shading all the cells that have the centers laying in the cone of shadow of any cell with a negative value $\mathcal{H}$; this is competitive with the first method in terms of computation time only at a low scale discretization.

The approach adopted for producing all the probability maps presented in this study is instead based on simple and local connection controls on the single cells of the grid: this in general shades less cells than the previous methods based on global radial lines, but it is much convenient in terms of computation time and gives results still consistent with 3D simulations. In particular the cells are checked following a precise ordering, starting from the center and considering increasing square frames; if $\mathcal{H}(i, j) > 0$ then the cell (i, j) is invaded only if one at least of the 1-2 adjacent cells that are crossed by the line from the vent to the center of (i, j) have been invaded (the strict ordering implies that in all the cases these two cells have just been checked). It is also possible to consider more adjacent cells: the 2-3 cells that have the center closer to the vent than the center of (i, j) or even all the 3-4 adjacent cells than are crossed by a line from the vent to any of the points of the cell (i, j). The first choice is preferred because of the possibility to propagate the invasion from more adjacent cells would imply an increased capability to turn around obstacles that is not specific of PDCs.

In some particular scenarios this integrated model have been compared with the 3D multi-phase simulations: the box model volumes that produce the best fit of the invaded area are about 5-10 times bigger than the correspondent 3D volumes; the motivation can be found in the fact that a continuous flow have been approximated with a transient collapse. Moreover, even the axisymmetric deposits of past eruptions are believed to be the sum of single members propagated in different directions, each involving less mass than a single big axisymmentric collapse. More accurate (direct) PDC propagation emulators capable of considering a partial overcoming of topographical barriers and blocking effects may be a fundamental step ahead.

Inversion problem Assuming to have sampled an area extension of y km^2 and a vent location x, the inversion problem corresponds to find the right initial conditions for invading that areal size from such source location. This approach strongly depends on the past deposits estimation: indeed the invasion model and the topographical barriers determine only the shape of the invaded area, its size being imposed from past data information. The direct model formal representation is the function $\sigma(\tilde{F})$ of Definition 3.8, that once fixed the vent location x (and the physical parameters of the flow) gives a relation between the maximum run-out l_{max} without topography and the area y invaded: $\sigma(\tilde{F})(x, l_{max}) \subseteq B$. An inversion problem solution is an l^*_{max} such that $G(l^*_{max}) = y$, where

$$G(l^*_{max}) := \int_B 1_{\sigma(\tilde{F})(x,l^*_{max})}(s)ds.$$

The function G in not even continuous, but it is increasing: using this property it is possible to use the secant method (Regula Falsi) to solve the inversion problem. Such method needs two starting values, one above and one below the solution l^*_{max}. To find the first one it is explicitly solved the inversion problem in the simpler case of the energy line model: because of the linearity of the decaying function it is equivalent to a single direct propagation in terms of computation time. Then it is iteratively sought a second initial point and finally it is implemented the secant method with two stopping conditions: one primary condition on the relative error of the area invaded $|G(l_{max}) - y|/y$, and a secondary condition on the step $|l^{n+1}_{max} - l^n_{max}|$ of the approximation sequence. This second condition can produce inaccuracy on the invaded area, but it is fundamental to stop the algorithm in the case the target area lays in a discontinuity of the function and it is impossible to reach precisely. Anyways it was directly tested that the inversion algorithm gives relative errors below 0.05 in more of the 95% of the extractions. The whole algorithm (reported in Chapter 6) needs about 5-6 iterations of the direct propagation method and it is fast enough to be run in the Monte Carlo simulation aimed at exploring on the PDC invasion maps the effects of the uncertainties affecting vent location and eruption scale. Alternative approaches for avoiding this inversion have been considered, but volume or mass estimates of past PDCs are very uncertain and difficult to obtain.

Monte Carlo structure Recalling notation of Definition 3.4, the mean $p(z)$ of the PDC invasion probability of point $z \in B$ with respect to all the random variables both aleatoric and epistemic, represents an information of the main importance. Fixed a vent location x and an areal size y, a single run of the PDC propagation emulator permits to obtain the indicator function of the invaded area: a function $P_{x,y} := 1_{F(x,y)}$ from B onto $[0, 1]$ taking only the two extreme values: $P_{x,y}(z) = 1$ if the point z is reached, $P_{x,y}(z) = 0$ otherwise. According to the doubly stochastic approach the average of such random indicator functions with respect to vent location, areal size and also the epistemic sources of uncertainty corresponds to calculate $E[P_{X,Y}(z)] = p(z) = E^E[[\check{p}(z)](\cdot)]$. Then, for quantifying the sensibility of the PDC invasion probability to the epistemic uncertainties involved, the structured Monte Carlo algorithm implemented permits to produce also a conditional PDC invasion probability map $[\check{p}(\cdot)](e)$ for each sample $e \in E$ of the sources of uncertainty. For each point $x \in D$ it is estimated the profile of the distribution of each variable $\check{p}(z)$ calculating an approximation of its percentiles (see also Figure 3.11). A first formula for the Monte Carlo implementation is

$$[\check{p}(z)](e) = E[1_{F(\check{X}(e,\cdot),\check{Y}(e,\cdot))}](z) \approx \sum_i^N 1_{F(\check{X}(e,\cdot),\check{Y}(e,\cdot))}(z)/N,$$

where $N \gg 1$, $\check{X}$ represents the vent location and $\check{Y}$ represents the PDC invaded area conditional on e. For approximating the profile of each $\check{p}(z)$ as a function of e it is produced a population of $N_2 \gg 1$ random functions $\check{p}(\xi)$, where ξ represents a random sample for epistemic uncertainty (as in Definition 1.1). It is adopted an implementation based on the discretization of the spatial distribution of vent opening location:

$$E[1_{F(\check{X}(e,\cdot),\check{Y}(e,\cdot))}](z) = E^{\mu_{\check{X}}(e) \otimes \nu_{\check{Y}}(e)}[1_{F(\cdot,\cdot)}](z),$$

and applying Fubini theorem for each $e \in E$ implies

$$= E^{\nu_{\check{Y}}(e)}\left[E^{\mu_{\check{X}}(e)}\left[1_{F(\cdot,\cdot)}\right]\right](z),$$

and discretizing on a grid of points $(x_j)_{j<n}$ for each $y \in \mathbb{R}_+$

$$E^{\mu_{\check{X}}(e)}[1_{F(\cdot,y)}] = \sum_j^n q_j(e)1_{F(x_j,y)},$$

where $q_j(e)$ are the discretized probabilities on the points of the grid. In conclusion the approximation adopted is

$$[\check{p}(z)](e) \approx \sum_i^{N_1} \sum_j^n q_j 1_{F(x_j,\check{Y}(e,\cdot))}(z)/N_1.$$

The total number of PDC simulations using the emulator is $N_2 \times N_1 \times n \approx 75 \times 15 \times 500 \approx 5 \times 10^5$.

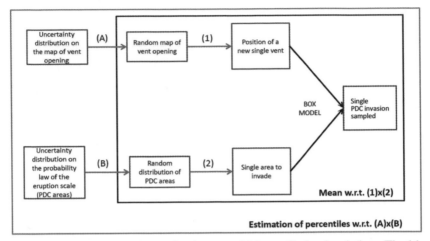

Figure 3.11. General scheme for the nested Monte Carlo simulation. The blue arrows represent the random samples which determine the elements in the small blue boxes, the pair of black arrows represents the box model PDC propagation. For obtaining each single PDC invasion map conditional on the epistemic assumptions sampled by (A) and (B) we calculated the mean of the PDC invasions (in the small red box) with respect to (1) and (2). The large black box envelopes this part of the scheme. The larger blue box envelopes the whole of the scheme, needed for calculating the uncertainty percentiles of the PDC invasion maps depending on the epistemic assumptions sampled by (A) and (B).

Remark 3.9. The output maps are the bi-dimensional plots of the mean probability of invasion $p(z) = E^E[[\check{p}(z)](\cdot)]$ and of the uncertainty bound functions p_1 and p_2, that are the 5th and 95th percentiles functions of the variables $[\check{p}(z)](\cdot)$ for each $z \in B$. That maps are obtained working on a discrete grid of cells 500 m aside, with tests of sensibility assuming a grid of cells 250 m aside. Assuming to choose a threshold probability ϵ, there are two different approaches for defining the points with probability to be invaded $p > \epsilon$ in some sense. A first straightforward way is to take the contour ϵ on the map of p, but this means to consider the epistemic uncertainty in the same way of the physical (aleatoric) variabilities of the phenomena. Another approach is instead to take the contour ϵ of the map of p_2: the 95th percentiles functions with respect to the uncertainties involved; this means that it is assumed to take almost the worst outcome following the epistemic uncertainty. Even more detailed approaches may be followed considering also the map of p_1.

Chapter 4
Time-space model for the next eruption

4.1. Summary

The main objective of this chapter is developing a robust temporal model capable of producing a background (long-term) probability distribution for the time of the next explosive eruption at Campi Flegrei. In the last 15 ka, intense and mostly explosive volcanism has occurred within and along the boundaries of the caldera (see Chapter 2; *e.g.* [97]). Eruptions occurred closely spaced in time, over periods from a few centuries to a few millennia, and were alternated by periods of quiescence lasting even several millennia; sometimes events also occurred closely in space thus generating a cluster of events (*e.g.* [96]). As a consequence, activity has been generally subdivided into three distinct epochs, *i.e.* Epoch I, 15 - 10.6 ka; Epoch II, 9.6 - 9.1 ka, and Epoch III, 5.5 - 3.8 ka BP (*e.g.* [122, 145]). The most recent Monte Nuovo eruption (*e.g.* [56, 58, 79]) occurred in 1538 AD after more than 2.7 ka from the previous one. Unfortunately, there is a remarkable epistemic uncertainty on the eruptive record, affecting the time of eruptions, location of vents as well as the erupted volume estimates. Other studies including information about the time space eruptive behaviour of Campi Flegrei are [133, 139, 57, 54].

The temporal model will focus on the eruptive clusters recognition and uncertainty quantification; it aims at being informative also about the unknown long term consequences of the Monte Nuovo eruption, which at the moment seems to be an outlier in the temporal record. In particular the first purpose of the study is the modelling of the epistemic uncertainty by using a quantitative probabilistic approach and obtaining estimates about the temporal and spatial distribution of the volcanism. It is adopted a time-space doubly stochastic Poisson-type model with a local self-excitement feature able to re-produce clustering events that are consistent with the reconstructed observed pattern at Campi Flegrei.

Our results allow to estimate the temporal eruptive base-rate of the caldera as well as its capacity to generate clusters of events, under different volcanological assumptions. The analysis allows also to discriminate between the initial and the main part of the eruptive epochs as well as to consider separately the different behaviour of the eastern and western sectors of the caldera. The results from Epoch I record give a rate of generation of new clusters of one on 148 years, while for Epoch III we estimated one new cluster each 106 years in average. The duration of the self-excitement appears quite different between the epochs, with an average duration of 658 years for Epoch I and about 96 and 101 years respectively for Epoch II and Epoch III. Considering only the first parts of the eruptive epochs the rate of generation of new clusters is still around one on 150 years, although the clusters tend to last longer. Considering separately the events of the western sector produces one new cluster each 400 - 450 years, but the clustering features are the weakest and the most of the single events must be assumed as independent clusters.

The analysis of the past record implies that the Monte Nuovo event is too distant in time from the end of the Epoch III for being considered a continuation of it: a repose time of ~3 ka is potentially longer than the periods of quiescence that separate the epochs. In particular with the base eruption rate obtained from the rest of Epoch III the probability of producing such a long repose time is below 10^{-12}. Moreover, the past epochs volcanic activity seems to have started in the western sector of Campi Flegrei caldera: the possibility that a new eruptive epoch began 477 years ago cannot be neglected, and an epoch of activity containing only Monte Nuovo event would represent a behaviour never observed in the previous 15 ka. If we assume Monte Nuovo as the first event of a new eruptive epoch, the estimates about the next eruption time are of 103 years in average based on all the past activity, and of 470 years relying only on the western sector eruptive record. In both cases the physical variability is large: 5th and 95th probability percentiles range from 1/20 of these estimates to 3 times, and are respectively [5, 318] years and [25, 1467] years. The considered epistemic uncertainty affecting all the estimates is relevant, and have been quantified as ±30% and ±35% respectively, slightly skewed on the positive values.

4.2. Methodology

Various preliminary analyses and figures, are aimed at making more evident the presence of clusters, including the not negligible effects of the main sources of epistemic uncertainty. The choice of the Hawkes counting processes (see also Chapter 5) has the purpose to re-produce the clusters of eruptions in time-space, and we develop a doubly stochastic generalization of such processes called Cox-Hawkes processes for implementing the epistemic uncertainty. Then we obtain some estimates for the next eruption time based on this probability model.

The simplest choice of counting process would be a homogeneous Poisson process; however it is a too crude approximation of reality, based on the remarks of the following sections, where we emphasized the non-homogeneity in time (and space) of volcanic events at CF. A homogeneous Poisson process assumes that the waiting times between two events are independent identically distributed exponential random variables; the inverse of the average value of such waiting times is called the intensity λ of the process. As a consequence the intensity function λ has also the meaning of the average density of eruptive events: the integral $\int_{t_0}^{t_1} \lambda dt$ is the average number of events in the time interval $[t_0, t_1]$.

The next simplest choice would then be a non-homogeneous Poisson process, which corresponds to take an intensity $\lambda(t)$ that changes as a function of time: it may be possible to produce clusters first increasing and then decreasing the intensity function, and such processes are easy to simulate from homogeneous processes conditionally sampled (e.g. [46, 87] and Definition 5.36). An approach involving non-homogeneous processes may assume to fit a $\lambda(t)$ directly on the eruption rate observed during one of the epochs, but such model could be over-fitted over past data: the three epochs have different durations and patterns and even mixing them risks to rely too much on the past behaviour.

To solve this problem, the approach we have adopted is to assume a self-excitement of the intensity function of a ground homogeneous process: a constant base rate produces the background events, each of them having the chance of spreading an offspring, and even each of these spawn events has a chance to produce offspring. This is called a branching process and the families of offspring events can replicate clustering phenomena with a random pattern (see also Chapter 5). The intensity $\lambda(t)$ is defined as the sum of a constant term λ_0 and of a time dependent random (positive) term λ_{cl} that represents additional intensity produced by each event for a prescribed time range after its occurrence. In particular the process representing the number of eruptive events that occurred in each zone of the caldera as a function of time is taken in the class of multivariate Hawkes processes; e.g. [44, 46, 49] for more information, and [15] for another application in volcanology.

A Hawkes process Z behaves as follows: the time rate of new events is, at time t,

$$\lambda(t) = \sum_{k \geq 1} \varphi(t - T_k) 1_{\{T_k \leq t\}},$$

where $(T_k)_{k \geq 1}$ are the event times for Z, and φ is a positive function representing self-excitement decay. In other words, each time Z has a jump, it excites itself in that it increases its rate of jump. If φ decreases to 0 (the usual case, *e.g.* an exponential function with negative exponent or a sigmoid function) then the influence of a jump decreases and tends to 0 as time evolves.

This study included two phases: first the volcanological information is reported and a quantitative model of epistemic uncertainty affecting past eruption data is constructed, then the family of counting processes representing the number of vents (or fissures) opening in each zone as a function of time it is defined. In particular in the first phase the times, locations and volumes of dense rock equivalent (VDRE) erupted are randomly sampled and estimated through Monte Carlo simulations. Then it is presented a qualitative description for the main features of time-space structure of past activity, and also through the probabilistic simulation of the local and global erupted volumes as a function of time is remarked the presence of local clusters of events and in general of a recurrent behaviour of the volcanic system. In the second phase, the main parameters of the Hawkes process are fitted on the record of past activity, incorporating the effect of the sources of epistemic uncertainty. We focus on the case of processes with self-excitement locally in space, without assuming interaction between different zones. Finally, through another Monte Carlo simulation several estimations are developed about the time remaining before the next eruption at Campi Flegrei, based on conditional sampling and diverse volcanological assumptions.

4.3. The probability model for epistemic uncertainty on the past record

The available volcanological record about past eruptive events (see [122, 145]) includes four types of information: the uncertain location of the eruptive vents/fissures; the ordered stratigraphic sequence of the eruptions (unsure in a few cases); some large dating time windows for a subfamily of the eruptions, *i.e.* 2.5[th] and 97.5[th] percentiles; VDRE estimates for the eruptions, some of them affected by a large uncertainty.

The spatial localization of past vents relies on the new data of Chapter 2: the uniform ellipses of uncertainty and the caldera partitioning constructed are adopted (Figure 4.1 and also Figure 2.2-2.3). Each eruption has been associated with an element of the partition if its ellipse does not spread on more zones, or otherwise randomly sampled during the Monte Carlo simulation in proportion of the fraction of ellipse contained in each zone. In addition we define a separation line between the western and the eastern sectors of the on-land portion of Campi Flegrei caldera, corresponding respectively to the zones 1-5 and 6-13 accordingly to the partition of Figure 4.1.

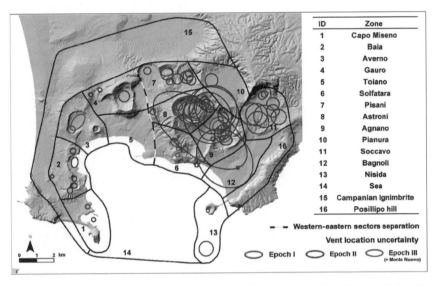

Figure 4.1. Partitioning of the caldera in 16 zones (see also Figure 2.4). The colours of the ellipses correspond to the epoch of activity which belong the events. The black-and-yellow dashed line separates eastern and western sectors.

The eruptions pattern of Epoch I consists of several events on the boundary of Neapolitan Yellow Tuff collapse, with evident spatial clusters in the zones of Soccavo and Pisani; in addition there was a remarkable activity in the central-eastern sector, but the relevant spatial uncertainty affecting it makes very difficult the recognition of clusters. The eruptions of the western sector seem less clustered and are much aligned on the caldera rim. During Epoch II five of the eight eruptions concentred in the central eastern part of the caldera, maybe on the boundary of a previous collapse; activity on the western sector was also present in the zone of Baia. The eruptions pattern of Epoch III is much more centralized than the previous: there are three quite evident spatial clusters centered respectively in Agnano, Astroni and Solfatara zones involving all the eruptions except for 5 that were in proximity of the caldera rim. A preliminary analysis of clusters of Epoch III is detailed in the following, taking into account also the time sequence.

The stratigraphic sequence and the time windows for past events, reported in Table 4.1, largely relies on [145] with a few minor modifications and updates due to the most recent research findings and uncertainty assessments (see also Chapter 2 and [17]): in particular in Epoch I the new eruptions of S3s4-1 and S3s4-2 have been added and Archiaverno event was eliminated from the record; in Epoch II the new eruptions of Baia and Monte Spina lava dome were considered; in Epoch III Solfatara lava dome and Olibano tephra have been added. Moreover the stratigraphic order of some Epoch III eruptions was modified in the initial part of the epoch and the first events after Agnano Monte Spina.

EPOCH I

ID	Name	Time [a] 2.5%ile	Time [a] 97.5%ile	VDRE [hm3] 5%ile		VDRE [hm3] 95%ile	E/W	Zone
1	Bellavista	-	-	10	-	100	W	2
2	Mofete	-	-	10	-	100	W	2
3	Gauro	12721	15511	250	500	750	W	4
4	Santa Teresa	-	-	10	-	100	E	12
5	La Pietra	-	-	10	-	100	E	6
6	La Pigna 1	12749	13110	10	-	100	E	11
7	La Pigna 2	-	-	10	-	100	E	11
8	Torre Cappella	-	-	10	-	100	W	2
9	Minopoli 1	-	-	10	-	100	E	11
10	Paradiso	-	-	10	-	100	E	11
11	Soccavo 1	-	-	250	500	750	E	11
12	Gaiola	-	-	10	-	100	E	13
13	Pomici Principali	11915	12158	425	850	1275	E	8-9-10
14	Paleo Pisani 1	-	-	10	-	100	E	7
15	Paleo Pisani 2	-	-	100	-	300	E	7
16	Soccavo 2	-	-	10	-	100	E	11
17	Soccavo 3	-	-	10	-	100	E	10
18	S4s3_1	-	-	100	-	300	E	8-9-10
19	S4s3_2	-	-	10	-	100	E	8-9-10
20	Soccavo 4	-	-	100	-	300	E	11
21	Paleo San Martino	-	-	10	-	100	E	7
22	Minopoli 2	-	-	10	-	100	E	11
23	Soccavo 5	-	-	10	-	100	E	11
24	Pisani 1	-	-	100	-	300	E	7
25	Pisani 2	-	-	100	-	300	E	7
26	Fondo Riccio	-	-	0	-	10	W	4
27	Concola	-	-	0	-	10	W	4
28	Montagna Spaccata	-	-	10	20	30	E	7
29	Pignatiello 1	-	-	10	-	100	E	6-7-8-9-12
30	Pisani 3	10516	10755	10	-	100	E	7
31	Casale	-	-	10	-	100	E	6-9-12
32	Bacoli	11511	14154	100	200	300	W	1
33	Porto Miseno	10347	12860	10	-	100	W	1

EPOCH II

ID	Name	Time [a] 2.5%ile	Time [a] 97.5%ile	VDRE [hm3] 5%ile		VDRE [hm3] 95%ile	E/W	Zone
1	Baia	-	-	0	-	10	W	2
2	Fondi di Baia	9525	9695	20	40	60	W	2
3	Sartania 1	9500	9654	10	-	100	E	8
4	Monte Spina lava dome	-	-	0	-	10	E	9
5	Costa San domenico	-	-	10	-	100	E	9
6	Pigna San Nicola	9201	9533	100	-	300	E	9
7	Sartania 2	-	-	10	-	100	E	8
8	San Martino	9026	9370	25	50	75	E	7

EPOCH III

ID	Name	Time [a] 2.5%ile	Time [a] 97.5%ile	VDRE [hm3] 5%ile		VDRE [hm3] 95%ile	E/W	Zone
1	Agnano 1	5266	5628	10	20	30	E	9
2	Agnano 2	-	-	5	10	15	E	9
3	Averno 1	5064	5431	10	-	100	W	3
4	Agnano 3	-	-	95	190	285	E	9
5	Cigliano	*	*	25	50	75	E	8
6	Pigniatiello 2	*	*	10	20	30	E	9
7	Capo Miseno	3259	4286	10	20	30	W	1
8	Monte Sant'Angelo	4832	5010	100	-	300	E	9
9	Paleoastroni 1	4745	4834	25	50	75	E	8
10	Paleoastroni 2	4712	4757	100	-	300	E	8
11	Agnano Monte Spina	4482	4625	425	850	1275	E	8-9
12	St. Maria delle Grazie	4382	4509	10	-	100	E	6
13	Olibano lava dome	*	*	0	-	10	E	6
14	Paleoastroni 3	-	-	10	20	30	E	8
15	Solfatara lava dome	*	*	0	-	10	E	6
16	Olibano tephra	*	*	10	-	100	E	6
17	Accademia lava dome	-	-	0	-	10	E	6
18	Solfatara	4181	4386	15	30	45	E	6
19	Averno 2	**	**	35	70	105	W	3
20	Astroni 1	4153	4345	30	60	90	E	8
21	Astroni 2	-	-	10	20	30	E	8
22	Astroni 3	-	-	80	160	240	E	8
23	Astroni 4	-	-	70	140	210	E	8
24	Astroni 5	-	-	50	100	150	E	8
25	Astroni 6	-	-	60	120	180	E	8
26	Astroni 7	4098	4297	35	70	105	E	8
27	Fossa Lupara	3978	4192	10	20	30	E	7
28	Nisida	3213	4188	10	20	30	E	13

Table 4.1. Record of times, erupted VDRE and locations (eastern or western sectors and partition zones of Figure 4.1) of the events at Campi Flegrei, with uncertainty bounds. The events with unknown ordered sequence are reported in green, the events with both ordered sequence and datation are reported in red. The VDRE estimates possessing estimated values in [145] are reported in blue. In Epoch III dataset, the pairs of events with uncertain order are indicated with *, the two simultaneous events with **. Data modified from [145].

The following is the formal definition of the probability model for epistemic uncertainty that samples the ordering the times and the locations of the past eruptions: technical details on the formal construction of these random variables are in Appendix A of this chapter. It partially relies on the notation of Definition 1.1.

Definition 4.1 (Time-space record with uncertainty). Let $(w_i)_{i=1,...,n}$ be the set of all the eruptive events considered. Assume that τ is a random variable from $(E, \mathcal{E}, \eta)$ to the space $\mathcal{S}(n)$ of the permutations of $\{1, \ldots, n\}$ such that $(v_j)_{j=1,...,n}$, where

$$v_j := w_{\tau(j)}, \quad \forall j,$$

represents a random sample for the ordered family of eruptive events. Let $(t_j)_{j=1,...,n}$ be a vector of real random variables from $(E, \mathcal{E}, \eta)$ to $\mathbb{R}^n_+$, each t_j representing the time of eruptive event v_j, consistent with the datation bounds available. For each $j = 1, \ldots, n$ let V_j be a random variable from $(E, \mathcal{E}, \eta)$ to $(A, \mathcal{B}(A))$ representing the location of the eruption v_j. We define the random set of random variables

$$\Theta_l =: \{t_j \ : \ V_j \in A_l\}$$

representing the times of each eruption v_j that occurred in the zone A_l. We adopt the notation $\Theta_l = (t_j^l)_{j=1,...,n_l}$.

The production of the probability distribution of past eruptions time is assessed by a Monte Carlo simulation, based on simple conditional sampling procedures (see also [15]), following three main steps.

1. At first the uncertain orders of some pairs of events are randomly sampled, including also the constrain of two simultaneous events (see Chapter 2 and [98]); this was done only for some eruptions of Epoch III due to lack of detail in the previous epochs. In particular Averno 2 and Solfatara were assumed contemporaneous, while for Pignatiello 2 and Cigliano, Olibano lava dome and Santa Maria delle Grazie, Olibano tephra and Solfatara lava dome we assumed a randomized order: the first pair with 50% and the other pair with 25% probability to have a different order than in Table 4.1. Anyways this has a negligible effect on the uncertainty model output because none of these was a very large event and some of these pairs affected the same zone of the caldera.

2. After this preliminary phase, the times for the eruptions associated with dating time windows are then sampled, assuming symmetric triangular probability distributions with the percentiles shown in Table 4.1, and repeating the samples that violate the stratigraphic order except for three events of the western sector, assumed free to change their place in the eruptive sequence. In particular Bacoli and Porto

Miseno datation ranges in Epoch I and also Capo Miseno range in Epoch III are much larger than the constraints coming from the stratigraphic sequence, which is very uncertain for them. It is remarkable that also Gauro in Epoch I and Nisida in Epoch III are affected by a very large epistemic uncertainty compared to the others, but in both cases it was consistent with the eruptive sequence. The available datation ranges interest a total of 23 eruptive events: 6 of Epoch I, 4 of Epoch II and 13 of Epoch III; this information comes from [145] with some modifications: in Epoch I the uncertainty range of Archiaverno event was not considered because the eruption was eliminated from the record, and the datation range assumed for Astroni 3 was assigned instead to Astroni 1 (R. Isaia, personal communication).

3. The remaining subsequences of eruptions between the times just fixed, are sampled as ordered families of independent uniformly distributed random variables. To give an example, assume that Astroni 1 and Astroni 7 eruption times have been sampled at $t_1 > t_7$ years BP: then five events are sampled uniformly and independently in the interval between that times obtaining $t_2 > t_3 > t_4 > t_5 > t_6$ years BP; hence the first time is assigned to Astroni 2, the second to Astroni 3 and so on, respecting the known ordered sequence. We remark that this uniform sampling could partially hide a potentially stronger clustering behaviour, in particular during Epochs I and II. In addition, for such epochs a few eruptions starting and ending the stratigraphic sequence are without informative datation ranges, so their times were sampled for simplicity a number of years uniformly distributed between 0 and 100 after the previous and before the successive: we chose this time scale because it is representative of the average time interval that separates the events during the eruptive epochs. *e.g.* if Gauro is sampled at t years BP, then Mofete and Bellavista are assumed $t + u_1$ and $t + u_1 + u_2$ years BP respectively, where u_1 and u_2 are uniform samples in [0,100] years.

The erupted volumes (VDRE) also rely on [145]: some of the eruptions have a single valued estimate, and for them it was assumed a triangular sampling with 5th and 95th percentiles corresponding to ±50% relative errors, coherently with [94]. The most of the other eruptions possess only inequality bounds: they have been uniformly sampled inside three separate intervals associated to different volume sizes: $[0 - 0.01]$ km^3 for very small eruptions/lava domes, $[0.01 - 0.1]$ km^3 for medium eruptions, $[0.1 - 0.3]$ km^3 for larger eruptions. The minor eruptions of Concola and Fondo Riccio and the newly included eruptions of Solfatara lava dome,

Baia and Monte Spina lava dome were assumed as small sized, S4s3-2 and Olibano tephra as medium sized and S4s3-1 as large sized (R. Isaia, personal communication).

The most large erupted volumes correspond to events located in the eastern sector of the caldera, with the only exceptions of Gauro and Bacoli during Epoch I. There were four major eruptions overcoming 0.5 km^3 VDRE: Gauro, Soccavo 1, Pomici Principali during Epoch I, and Agnano Monte Spina during Epoch III. Moreover, each of the most evident clusters seem to contain at least one large eruption, with the exception of Solfatara cluster in Epoch III. In particular, during Epoch I large eruptions were located in the zones of Gauro (Gauro volcano), Soccavo (Soccavo 1 and Soccavo 4), the central eastern zones (Pomici Principali, S4s31), Pisani (Paleo Pisani 2, Pisani 1, Pisani 2), and Capo Miseno (Bacoli). During Epoch II the only large eruption was in Agnano zone (Pigna San Nicola). Epoch III presented large eruptions only in the zones of Agnano (Agnano 3, Monte Sant'Angelo, Agnano Monte Spina) and Astroni (Paleoastroni 2, Astroni 3-4-5-6). An analysis of the localized erupted volumes as a function of time is detailed in the sequel, and some examples are plotted in Figure 4.6.

4.3.1. Detailed exploration of data

Accordingly to the probability model defined, the average duration of eruptive epochs and periods of quiescence is reported with its 5th and 95th epistemic uncertainty percentiles by a simple Monte Carlo simulation: Epoch I [2.56 - 3.63 - 4.74] ka, the first period of quiescence [0.79 - 0.95 - 1.10] ka, Epoch II [0.29 - 0.46 - 0.63] ka, the second period of quiescence [3.53 - 3.75 - 3.97] ka, Epoch III [1.37 - 1.81 - 2.24] ka; the time interval between Nisida and Monte Nuovo depends directly on the uncertainty affecting the first: [2.79 - 3.22 - 3.65] ka, and the latter occurred 477 years BP. The three epochs have remarkably different durations: Epoch I lasted about twice Epoch III, while Epoch II was much shorter than both and lasted a quarter of Epoch III duration. The first period of quiescence is much shorter than the second, its duration is even lower than the range of uncertainty affecting the times of some eruptions in the first part of Epoch I; anyways the first period of quiescence is twice the time interval between Monte Nuovo eruption and the present time (year 2015).

Another remarkable observation is that the period of quiescence of 3.22 ka in mean between Nisida and Monte Nuovo is more than 3 times longer than the first period of quiescence between Epoch I and Epoch II lasted 0.95 ka in mean, and it is very similar to the long period of quiescence between Epoch II and Epoch III, of 3.75 ka in mean. Hence the assumption to consider Monte Nuovo as a continuation of the Epoch III activity is not consistent with the separation in epochs. If it is assumed the division in epochs of activity and periods of quiescence, basing only on time records information the only reasonable possibility is that Monte Nuovo opened a new epoch; what it is uncertain is the duration and the number of event of such epoch of activity, that could either contain the Monte Nuovo eruption alone or develop in the future a pattern similar to one observed during past epochs. However an eruptive epoch containing only one event would represent a behaviour never observed in the previous 15 ka of activity.

Figure 4.2 shows two samples of eruptions times and zones during Epoch I and III, the presence of clustering phenomena is quite clear. In Epoch I it is possible to recognize some hypothetical clusters: the sequences of Soccavo zone, Pisani and Astroni-Agnano present clear evidence of clustering. Also Epoch II may have presented very small clusters of couples or triplets of eruptions (not shown in figure) but in both cases the epistemic uncertainty overwhelms any more detailed description.

During Epoch III, by visual inspection of Figure 4.2 it is natural to identify three space clusters in time-space plus a cloud of reasonably homogeneous background points. The first cluster (in order of time) is made of the eruptions denoted by Agnano 1, Agnano 2, Agnano 3, Pignatiello 2, Monte Sant Angelo; the second one is made of the eruptions denoted by Santa Maria delle Grazie, Olibano lava dome, Solfatara lava dome, Olibano tephra, Accademia lava dome, Solfatara; the third one by Astroni 1 - 7.

Even this first analysis is a source of interesting discussion: the Paleoastroni sequence (possibly including Cigliano event and/or AMS) of eruptions may belong to an enlarged Astroni cluster or may also constitute a small separate cluster, the eruption of AMS may be included in the Agnano cluster (or in the Astroni cluster, if enlarged to include Paleoastroni) even though it is more distant in space and time from the others, and may also arise an issue about this first cluster to be formed instead by more smaller detached clusters; and everything is made worse to decide because of the uncertainty affecting it.

Then in Figure 4.3 is reported the eruptions number as a function of time with uncertainty. Epoch I and even the smaller Epoch II present some evident features despite the large uncertainties that affect them. An intensification of the activity rate is quite evident during Epoch I (in this case the rate change may correspond to Pomici Principali). The data of Epoch III are relatively more precise and it is easier to investigate the

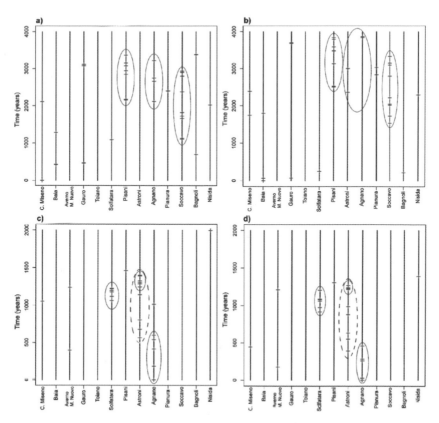

Figure 4.2. Random samples of eruption times and zones during Epoch I (a-b) and Epoch II (c-d). Each coloured dash represents an event, accordingly to the model for assessing the epistemic uncertainty. The red ellipses remark some qualitatively recognizable time-space clusters of activity; the black dashed ellipse remarks an hypothetically enlarged Astroni cluster including Paleoastroni events.

time-space structure of the eruptions record. Also in the second part of Epoch III after a relatively slow start it is possible to see an increase of the activity frequency, and the change in eruption rate seems to coincide with the climactic eruption of Agnano Monte Spina (AMS).

In particular may be identified two different stages, like two sub-epochs, separating 11 known eruptions before and 17 after to AMS; the time-structure is quite different, but the vent locations during these two periods affected similar and overlapping zones. In addition a slow-down of the activity is quite evident at the end of the epoch. These features are perhaps accentuated by the large uncertainties affecting the starting and the ending of both Epoch I and Epoch III. However the pictures seem not compatible with a homogeneous Poisson process and also the location in

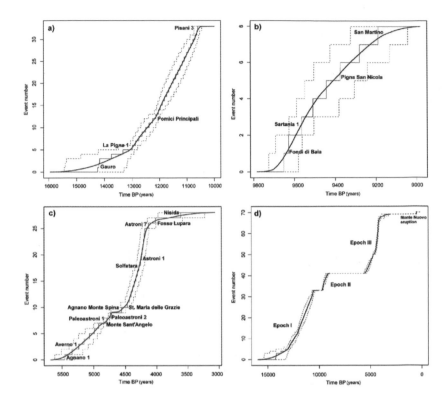

Figure 4.3. Event number as a function of time during Epoch I (a), Epoch II (b), Epoch III (c) and then during the entire record considered (including Monte Nuovo) (d), assuming the probability model described above. The bold line is the mean value, the narrow line is the 50th percentile and the dashed lines are 5th and 95th percentiles of the epistemic uncertainty. The labels correspond to the eruptions possessing datation bounds and ordering.

space of the eruptions is clearly non-homogeneous (Figure 4.1 and 4.2; see Chapter 5 for more information about Poisson processes).

We remark that this approach relies in principle on the counting of past eruptions in time space, but the number of the events itself could be a matter of discussion. Indeed it is widely accepted that Astroni activity was constituted by a sequence of 7 (or more) eruptions during 150 - 350 years (see [96]) and that the eruptive phases of AMS were instead much more concentred in time and they have been considered as a single event (see [53]). However the estimates of the cumulative erupted volume of Astroni sequence are quite similar to the volume estimates of AMS and analogue issues may arise on the phases of the Averno 2 activity or on the large eruptions of Epoch I. For making the probabilistic model results more robust and not only dependent on the event counting, the eruptive record includes some estimates of the erupted volumes, which are unfortunately affected by huge additional uncertainties because of the difficult procedure that must be followed to estimate them from the deposits. Such uncertainties are incorporated in the analysis, assessed by Monte Carlo samplings and it is still possible to observe the presence of clusters of events in time-space also on the rates of eruptive volumes.

In Figure 4.4 are shown the cumulative erupted volumes: it is worth noting the different scale of Epoch II, 0.5 ± 0.1 km^3, involving volumes 5 times smaller than Epoch III, 2.6 ± 0.5 km^3, and even 8 times smaller than Epoch I, 4.2 ± 0.7 km^3.

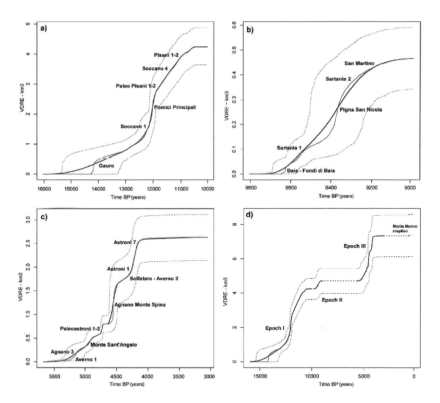

Figure 4.4. Cumulative volume erupted as a function of time during Epoch I (a), Epoch II (b), Epoch III (c) and then during the entire record considered (including Monte Nuovo) (d), assuming the probability model described above. The bold line is the mean value, the narrow line is the 50th percentile and the dashed lines are 5th and 95th percentiles of the epistemic uncertainty. The labels correspond to the largest eruptions of each epoch.

In Figure 4.5 the cumulative volumes erupted by the eastern part of the caldera are separated from the volumes from the western, which even including the significant activity of Gauro volcano in the Epoch I are remarkably smaller. It can be also noted that the activity in the western sector seems mainly associated to the initial phases of the eruptive epochs, compatibly with considering the recent Monte Nuovo eruption as the first one of a new epoch.

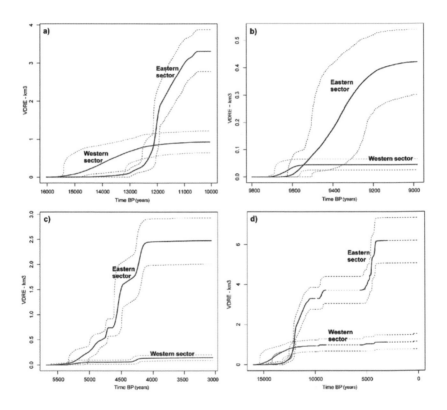

Figure 4.5. Cumulative volume erupted as a function of time during Epoch I (a), Epoch II (b), Epoch III (c) and then during the entire record considered (including Monte Nuovo) (d), with a separation between the eastern and the western sectors of the caldera. The bold line is the mean value and the dashed lines are 5th and 95th percentiles of the epistemic uncertainty.

Figure 4.6 shows the cumulative volumes locally erupted by the main zones affected by the larger or recurrent explosive activity during Epochs I and III: the plots clearly highlight the presence of clusters of events in time and space.

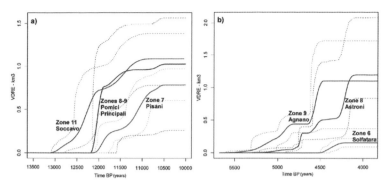

Figure 4.6. Localized cumulative volume erupted as a function of time during Epoch I (a) and Epoch III (b), from three different zones characterized by intense or recurrent activity. The bold line is the mean value and the dashed lines are 5th and 95th percentiles of the epistemic uncertainty.

4.4. The probability model for re-producing the eruption activity in time-space

We define a family of counting processes representing the number of vents opening in each zone of the caldera as a function of time. The model adopted relies on a 'Cox-Hawkes process', *i.e.* a doubly stochastic Hawkes process (see Chapter 5), including a spatial localization in the different sectors of the caldera. The Hawkes processes are non-homogeneous Poisson processes (NHPP) in which the intensity rate increases with a jump whenever an event occurs and instead decreases (often following exponential or sigmoid decay curves) as time passes without any event occurring. The Cox processes are simply the doubly stochastic version of the NHPP, in which the model parameters are assumed affected by uncertainty. The innovative model developed presents both these properties; in particular we explored the case of an exclusively local self-excitement, *i.e.* without interaction between different zones.

Definition 4.2 (The Cox-Hawkes process). Let $Z = (Z^l)_{l=1,...,N}$ be a doubly stochastic multivariate Hawkes process on $(\Omega, \mathcal{F}, P)$, adopting the non-trivial structure of Definition 1.1. Let φ be an application from E to the functional space of continuous decreasing functions on $\mathbb{R}_+$, representing the diminishing of self interaction for the process Z. For each $l = 1, ..., N$, let λ_0^l be a random variable on $(E, \mathcal{E})$ representing the base rate of the process Z^l. The intensity function of the component Z_l is then expressed by

$$\lambda^l(t, \omega) = \lambda_0^l(e) + \sum_{t_i^l(w) < t} [\varphi(e)](t - t_i^l(\omega))$$

$$= \lambda_0^l(e) + \int_0^t [\varphi(e)](t - u) dZ_u^l(\omega), \quad \forall l = 1, \ldots, N,$$

where we assume $e = \xi(\omega)$ of Definition 1.1.

For each zone j of the caldera, the self-interaction decay function φ is assumed exponential

$$\varphi(t) = h\exp(-kt),$$

where k and h are positive parameters. In addition, due to the epistemic uncertainty that affects the times and the locations of past events, such Hawkes processes have to be doubly stochastic in the sense that the parameters are randomly distributed as a function of the sources of uncertainty considered. Doubly stochastic NHPP are called Cox processes; e.g. [44, 46] for more information and [88, 89, 90] for examples of applications in volcanology. Each of the physical parameters of the model: base rate λ_0, time scale of excitement decay T and mean number of offspring events μ, is represented as a random variable on the epistemic space $(E, \mathcal{E}, \eta)$. A maximum likelihood (ML) procedure is implemented inside a Monte Carlo simulation to calculate the uncertainty distribution of the parameters as a function of the sources of epistemic uncertainty described above: time, sequence and location of past events.

Definition 4.3 (Conditional Cox-Hawkes processes). Let π^3 be a measurable function from $(W, \mathcal{W}, M)$ to the space of l-dimensional counting measures, representing the projection of the physical space onto the set of next eruptions times in each of the caldera zones. We assume that

$$Z(\omega) = \pi^3\left(\chi(\xi(\omega), \omega)\right), \text{ for almost every } \omega \in \Omega,$$

and we define the point process $\check{Z}$ from $(E \times \Omega, \mathcal{E} \otimes \mathcal{F}, \eta \otimes P)$ to $(A, \mathcal{B}(A))$ as

$$\check{Z}(e, \omega) := \pi^3\left(\chi(e, \omega)\right).$$

For each $e \in E$ the point process $\check{Z}(e, \cdot)$ on $(\Omega, \mathcal{F}, P)$ represents the set of next eruptions times at Campi Flegrei once adopted the epistemic assumption e.

More formally, the epistemic uncertainty of the Hawkes process Z comes from the random vectors $(\Theta_l)_{l=1,\ldots,N}$ of Definition 4.1 through the parameters of the random intensity functions λ^l.

Definition 4.4 (The parameters of the process). Let h and k be real positive random variables on the space $(E, \mathcal{E}, \eta)$; we define the self interaction function from the class of exponential functions:

$$[\varphi(e)](s) = h(e)\exp(-k(e)s).$$

Let λ_0 be a random variable on $(E, \mathcal{E})$ representing the base rate of $\sum_{l=1}^N Z_l$. Each one of local base rates is defined as $\lambda_0^l(e) = [\mu_{\check{X}}(e)](A_l)\lambda_0$, where for each $e \in E$, $\mu_{\check{X}}(e)$ is a conditional probability map of vent opening (see Definition 1.3).

A global base rate λ_0 is assumed for the whole caldera, and the local base rate of each zone is calculated in proportion to the frequency of the number of events observed in that location on the total; another possibility may be to use the spatial frequency during a particular epoch, but it might be a choice over-fitted to its specific spatial pattern, and it have been preferred to assume global parameters. Also the parameters k and h are defined independently of the zone, but in the sequel it is explored the possibility of assuming different parameters for the eastern and western sectors of the caldera.

In particular the parameter k is proportional to the time $T = k/ln(20)$ needed for the decay of the 95% of self-excitement (*i.e.* the integrated additional intensity on the times above T is 5% of the total); a smaller k produces concentrated and better separated clusters, while instead a larger k produces longer clusters, easily with overlapping durations. Moreover h permits to define the mean offspring $\mu = h/k$ of an eruption, *i.e.* the mean number of eruptions generated by the additional intensity caused by a single event; a larger h (compared to k) produces clusters including more events, with the critical threshold $h \leq k$ otherwise the average size of the clusters diverges to infinity.

The Galton-Watson representation of the process $\sum_l Z^l$ (see Chapter 5) permits to obtain the total base rate λ_0 as a function of the total number of past eruptive events n and the duration of the epochs of activity. The explicit definition of h and k as a function of $e \in E$ is based on a maximum likelihood procedure with respect to the random vectors $(\Theta_l)_{l=1,\ldots,N}$. Additional information on the likelihood expression is reported in Chapter 5. Some samples of the model outcomes are shown in Figure 4.7.

Proposition 4.5 (The likelihood expression). *Let $[0, t]$ be a closed interval, representing the time domain in consideration. The likelihood of the families $(\Theta_l)_{l=1,\ldots,N}$ being the instants of the points of the processes Z^l in $[0, t]$ for each $l = 1, \ldots, N$, assuming that $t \geq \max\left(\bigcup_l \Theta_l\right)$, is obtained by*

$$L\left((\Theta_l)_{l=1,\ldots,N}, t\right) = \prod_{l=1}^{N} L_l\left((t_i^l)_{i=1,\ldots,n^l}, t\right),$$

where for all $l = 1, \ldots, N$

$$L_l\left((t_i^l)_{i=1,\ldots,n^l}, t\right) = \left(\prod_{i=1}^{n^l} \lambda^l(t_i^l)\right) \exp\left(-\int_0^t \lambda^l(s)ds\right),$$

and

$$\int_0^t \lambda^l(s)ds = \lambda_0^l t - \sum_{i=1}^{n^l} \left(\frac{h}{k}\left(\exp(-k(t - t_i^l)) - 1\right)\right).$$

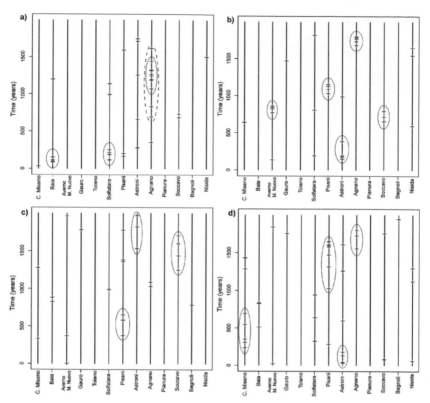

Figure 4.7. Four random samples of eruption times and zones obtained from the Hawkes model described above. The parameters adopted are the average ML values for the entire record of the three epochs. Each red dash represents an event. The blue ellipses remark some qualitatively recognizable time-space clusters of activity; the black dashed ellipse remarks an hypothetically enlarged cluster.

In case $\Theta_l = \emptyset$ for some l, we have $L_l(\emptyset) = \exp\left(-\int_0^t \lambda^l(s)ds\right) = \exp\left(-\lambda_0^l t\right)$.

It is easy to see that the likelihood is a real random variable on $(E, \mathcal{E})$, depending on the random vector $(\Theta_l)_{l=1,\dots,N}$. The log-likelihood of a time and space record can be easily expressed for each zone j as

$$\log L^j((t_i^j)_{i \leq N^j}, E) = \sum_{i=1}^{N^j} \log \lambda^j(t_i^j) - \lambda_0^j(E_2 - E_1) - N^j h/k$$

$$+ \sum_{i=1}^{N^j} \left(h/k \exp(-k(E_2 - t_i^j))\right),$$

where $E = [E_1, E_2]$ is the time interval considered (assumed to contain each t_i^j); the global log-likelihood of an eruptive record is the sum of the log-likelihoods of the single zones. Through a Monte Carlo simulation the ML parameters have been calculated for 2500 samples of the eruptive record, permitting also to estimate the uncertainty ranges for them (the implementation is reported in Chapter 6).

Remark 4.6. The maximizing procedure is repeated for each sample, finding the best fitting values for λ_0, h and k; with the purpose of reducing the dimension from three to two parameters and making it faster in the Monte Carlo simulation, the algorithm is implemented for finding directly T and n best fitting values, where $n := \lambda_0(E_2 - E_1)$ represents the average number of base rate eruptions that may be sampled on the considered time interval. The additional implicit relations $k = \ln(20)/T$, $h = k\mu \approx k(N-n)/N$ are imposed to close the algorithm; the last one is due to the branching structure of the process: each base rate eruption produces a cluster of $1/(1-\mu)$ eruptions in average, hence $N \approx n/(1-\mu)$ if it is assumed that each of the clusters is concluded inside the time interval considered (see also Chapter 5).

4.4.1. Conditional Monte Carlo for next eruption time estimation

The Cox-Hawkes process that maximizes the likelihood of the past data is also adopted for the forecast of the waiting time for the next eruption, in particular when a residual self-excitement from Monte Nuovo eruption is included. Excitement coming from previous eruptions is assumed negligible because of the long period of quiescence.

Definition 4.7 (The next eruption time distribution). Let Z_{mn} be a multivariate Cox-Hawkes process representing eruptions in each of the caldera zones, and starting from a situation without excitement except for the residual additional intensity from an event occurred $t_0 = 477$ years before time 0, in zone 3 (Averno-Monte Nuovo). Then define on $(\Omega, \mathcal{F}, P)$ the real positive random variable

$$Z^* := \min_l Z_{mn}^l,$$

representing the remaining time before the next eruption at Campi Flegrei. Let ϱ_{Z^*} be the probability measure that is the law of Z^* on $(\mathbb{R}_+, \mathcal{B}(\mathbb{R}_+))$; it is called a distribution of next eruption time.

Moreover it is possible to use the doubly stochastic structure (see Definition 1.1) of Z_{mn} and construct a conditional version of this variable on the epistemic uncertainty.

Definition 4.8 (Conditional next eruption time distribution). For each $e \in E$ let $\check{Z}_{mn}(e, \cdot)$ be a conditional multivariate Cox-Hawkes process representing eruptions in each of the caldera zones, and starting from a situation without excitement except that from an event occurred $t_0 = 477$ years before time 0, in zone 3. Then define on $(E \times \Omega, \mathcal{E} \otimes \mathcal{F}, \eta \otimes P)$ the real positive random variable

$$\check{Z}^*(e, \omega) := \min_l \check{Z}^l_{mn}(e, \omega),$$

and let $\varrho_{\check{Z}^*}$ be its law on $\mathbb{R}_+$. For each $e \in E$ the random variable $\check{Z}^*(e, \cdot)$ on $(\Omega, \mathcal{F}, P)$ represents the remaining time before the next eruption at Campi Flegrei once adopted the epistemic assumption e. Its law $\varrho_{\check{Z}^*}(e)$ is called probability distribution of next eruption time conditional on the epistemic assumption e.

The random variable $\check{Z}^*$ is assessed through a double Monte Carlo simulation with a nested structure, calculating a family of time samples depending on the different Cox-Hawkes model parameters obtained from each epistemic uncertainty sample. In particular once the uncertainty ranges for the parameters of the process have been estimated by a first Monte Carlo simulation, it is assessed a second Monte Carlo simulation for sampling the time remaining before the next eruption at Campi Flegrei, including Monte Nuovo residual self-excitement. This second simulation is based on independent triangular distributions that sample the stochastic process parameters. Additional research exploring the dependence properties between the model parameters λ_0, T and μ will be of the main importance for improving the uncertainty quantification.

A family of ML exponential distributions is then calculated for assessing the aleatoric variability of the model, conditional to each epistemic uncertainty sample. In Figure 4.8, we report the curves composed of mean values and of 5th and 95th percentiles of the probability density functions of such distributions. Furthermore, we express the mean and uncertainty bounds with respect of epistemic uncertainty of the mean and the 5th and 95th percentiles of the physical variability of the distributions. These procedure have been repeated under different volcanological assumptions for testing the sensitivity of the results. Also the effect of modifying the epistemic uncertainty with the Bayes theorem have been tested, concerning the 477 years passed without observing activity.

4.5. Results and discussion

In the following are first reported the parameters and the outcomes of the Cox-Hawkes process, then the discussion will focus on the next eruption time estimation results. In Table 4.2 are shown the ML values of the parameters $1/\lambda_0$, T, μ for each epoch, and also the results obtained maximizing the product of the likelihood of more epochs records. Similarly, Tables 4.3, 4.4, 4.5 report the ML values obtained under specific additional assumptions. Possible explorative/alternative features of the model may be related to the qualitative details highlighted about past activity: in particular the possible change of eruption rate with the climactic eruptions of Pomici Principali in Epoch I and Agnano Monte Spina in Epoch III, the differences between eastern and western parts of the system, and even the refusal of the division in eruptive epochs and periods of quiescence.

Other interesting information is obtained from the model: the probability of an eruption to generate offspring, the mean number and size of clusters of two or more events, and the probability distribution on the possible cluster sizes. Because of the branching structure of the process it is imposed that the chances of clusters sizes are a strictly decreasing sequence with respect to $\mathbb{N}$: there are many more smaller clusters than larger; an observed large cluster is more likely constituted by the superposition of two or three medium size independent clusters.

Assuming such simple model and Monte Nuovo eruption to be the starting event of a new eruptive epoch, the residual probability Q_{mn} of having in the future a second element of a cluster generated by it, and the likelihood L_{mn} of not observing any other eruption anywhere in the Campi Flegrei area for 477 years after it have been also calculated: both are significantly small; in the sequel it is explored if this might be a consequence of a different behaviour specific of the starting of the eruptive epochs or of the separation between eastern and western parts of the volcanic system.

Definition 4.9. Let Z be the Cox-Hawkes process representing (localized) eruptions starting from a situation without excitement except for an event at time zero in zone 3 (Averno-Monte Nuovo); let $t_0 = 477$. Then for each $e \in E$ we define the random variables

$$Q_{mn}(e) := P\left\{ \lim_{s \to +\infty} \tilde{Z}^3(e)_s > 0 \right\}, \qquad L_{mn}(e) := P\left\{ Z(e)_{t_0} = 0 \right\},$$

where $\tilde{Z}^3$ is a counting process with intensity $\lambda^3_{t+t_0} - \lambda^3_0$.

4.5.1. Parameters of the Cox-Hawkes process

In the following the epistemic uncertainty ranges of the variables, here represented as $(5^{\text{th}} \, perc - mean)/mean$ and $(95^{\text{th}} \, perc - mean)/mean$ will be included after the mean values. We remark that in general, the uncertainty ranges reported refer to the mean holding time of the new clusters $1/\lambda_0$ and not to the base rate λ.

The complete record of only eruptive epochs. In Table 4.2, the features of Epoch I are a base rate of one on 148 years, [-35%, +60%], (about 2/3 of the base rate of Epoch III), a self-excitement duration $T = 658$ years, [-90%,+100%], (6 times the duration of self-excitement in Epoch III), and a mean offspring $\mu = 0.30$ with an epistemic uncertainty range of [-55%, +40%]. The probability Q_{mn} of observing a second event of a cluster started by Monte Nuovo is 3.8% in average for Epoch I. With the assumption of Monte Nuovo as the initial event of a new eruptive epoch, the likelihood L_{mn} of no eruptions occurring in Campi Flegrei after Monte Nuovo for 477 years is [0.6%, 3.8%, 10.5%] based on Epoch I data (4 times greater than for Epoch III). Comparing the short Epoch II with the following Epoch III, we remark a base rate almost twice larger, a similar T with a larger uncertainty range, and a decreased μ meaning a less clustered structure. The probability Q_{mn} and the likelihood L_{mn} are both almost null based on Epoch II.

The ML parameters for Epoch III eruptive record correspond to a quite fast base rate of one on 106 years, [-25%,+35%], a self-excitement duration $T = 96$ years, [-90%, +105%], and a quite strong mean offspring $\mu = 0.42$, [-30%, +20%] that implies the generation of 5.5 clusters, [-35%, +40%], each composed of 3.2 elements in average, [-18%,+12%]. The probability Q_{mn} is almost null. The likelihood L_{mn} is only [0.2%, 0.9%, 2.3%] (average value with the 5th and 95th epistemic uncertainty percentiles) based on Epoch III data.

Reminding the description of past activity during Epoch III, there have been recognized three clusters of 5, 6 and 7 events respectively, plus 10 dispersed eruptions (see Figure 4.2). It is likely to not have recognized 1 to 3 couples of events from the background of single eruptions, and there is the chance of 1-2 of the clusters of being instead two smaller clusters superimposed: it is reasonable to estimate 4 to 8 clusters plus 4 to 8 single events; this corresponds to 10 to 14 events coming from the base rate (*i.e.* the sum of the clusters and the single events numbers that are consistent with the observations), and prescribing the estimated duration of 1.37 to 2.24 ka it is obtained a base rate from one on 224 to one on 98 years, that is compatible with the epistemic uncertainty range of the ML parameters calculated, although a bit lower.

The parameters maximizing the combined likelihood of Epochs I × II × III, *i.e.* of the three epochs assumed as independent samples, are much similar to Epoch III, except for the duration T that is doubled; consider-

COMPLETE ERUPTIVE EPOCHS

Statistics/Eruption record		Epoch I			Epoch II			Epoch III			Epochs I x II x III			Epochs I x III	
	5th %ile	mean	95th %ile	5th %ile	mean	95th %ile	5th %ile	mean	95th %ile	5th %ile	mean	95th %ile	5th %ile	mean	95th %ile
1/base rate (1/λ0) [a]	98	**148**	237	43	**63**	94	80	**106**	142	82	**105**	140	84	**112**	158
self excitement duration (T) [a]	60	**658**	1320	3	**101**	304	11	**96**	196	48	**189**	435	38	**230**	573
mean offspring (μ)	0.14	**0.30**	0.43	0.13	**0.23**	0.36	0.30	**0.42**	0.50	0.26	**0.41**	0.59	0.28	**0.45**	0.64
cluster generation probability	13%	**26%**	35%	12%	**21%**	30%	26%	**34%**	39%	23%	**34%**	44%	24%	**36%**	47%
number of clusters	2.4	**6.0**	9.9	0.6	**1.3**	2.1	3.6	**5.5**	7.6	-	**-**	-	-	**-**	-
mean cluster size	2.2	**2.7**	3.2	2.2	**2.4**	2.9	2.6	**3.2**	3.6	2.6	**3.1**	4.3	2.6	**3.3**	4.8
cluster sizes (1)	65.1%	**73.8%**	87.0%	69.6%	**79.3%**	88.3%	60.5%	**65.4%**	74.0%	55.5%	**66.2%**	76.8%	52.6%	**63.8%**	75.6%
probability distribution 2	10.5%	**16.6%**	18.2%	9.7%	**14.6%**	17.5%	16.5%	**18.1%**	18.4%	15.6%	**18.1%**	18.1%	16.0%	**18.3%**	17.7%
3	1.9%	**5.5%**	7.6%	1.6%	**4.0%**	6.7%	5.5%	**7.6%**	8.5%	4.7%	**7.4%**	8.9%	5.1%	**7.9%**	9.0%
4	0.4%	**2.2%**	3.8%	0.3%	**1.3%**	3.0%	2.2%	**3.8%**	4.5%	1.7%	**3.6%**	5.2%	1.9%	**4.0%**	5.4%
5	0.1%	**1.0%**	2.1%	0.1%	**0.5%**	1.5%	0.9%	**2.0%**	2.7%	0.7%	**1.9%**	3.3%	0.8%	**2.2%**	3.5%
6	0.0%	**0.4%**	1.2%	0.0%	**0.2%**	0.8%	0.4%	**1.1%**	1.7%	0.3%	**1.1%**	2.2%	0.3%	**1.3%**	2.5%
7	0.0%	**0.2%**	0.7%	0.0%	**0.1%**	0.4%	0.2%	**0.7%**	1.1%	0.1%	**0.6%**	1.6%	0.2%	**0.8%**	1.8%
8	0.0%	**0.1%**	0.5%	0.0%	**0.0%**	0.2%	0.1%	**0.4%**	0.8%	0.1%	**0.4%**	1.2%	0.1%	**0.5%**	1.4%
9	0.0%	**0.1%**	0.3%	0.0%	**0.0%**	0.1%	0.1%	**0.3%**	0.5%	0.0%	**0.2%**	0.9%	0.0%	**0.3%**	1.1%
10	0.0%	**0.0%**	0.2%	0.0%	**0.0%**	0.1%	0.0%	**0.2%**	0.4%	0.0%	**0.2%**	0.7%	0.0%	**0.2%**	0.8%
Probability of MN offspring after 477 years	0.0%	**3.8%**	10.1%	0.0%	**0.0%**	0.2%	0.0%	**0.0%**	0.0%	0.0%	**0.3%**	1.6%	0.0%	**0.9%**	3.7%
Probability of having 477 years with no events after MN	0.6%	**3.8%**	10.5%	0.0%	**0.1%**	0.5%	0.2%	**0.9%**	2.3%	0.2%	**0.8%**	2.2%	0.2%	**1.2%**	3.1%

Table 4.2. Maximizing the likelihood of Epoch I (first column), Epoch II (second column) or Epoch III records (third column), or also maximizing the product of the likelihoods of all the epochs records together (fourth column), or even excluding the shorter Epoch II (fifth column), we report several results. We include: the mean holding time of the new clusters $1/\lambda_0$, the duration of the self-excitement T, the mean offspring of each event μ (first three rows); the probability of cluster generation, the number of clusters of vents in case of single epochs, and the mean size of clusters (fourth to sixth rows); the discrete distribution of the cluster sizes (seventh row, composed of several lines); the probability Q_{mn} for an eruption of producing the first offspring event after 477 years (eighth row); the probability L_{mn} of passing 477 years after an eruption without observing other events anywhere in the caldera (ninth row). Mean values, 5th and 95th percentiles are reported, as a function of the sources of epistemic uncertainty considered.

ing Epochs I×III combined likelihood, implies an increase of T, with a minor increase of the base rate and of μ. The estimates for the average probability Q_{mn} obtained combining the epochs either with or without Epoch II are 0.3% and 0.9% respectively. The likelihood L_{mn} is similar to the one based on Epoch III (0.8% in average, increasing to 1.2% without considering Epoch II).

The first parts of the eruptive epochs. From past data it seems indeed that multiple and fast clusters tended to occur only in the second half of the longer epochs (I and III); for these reason the parameters and their epistemic uncertainty are also separately fitted on the first parts of the epochs (see Table 4.3). In particular it is remarkable that the seemingly similar features of Epoch Ia and Epoch IIIa bring different consequences on the ML parameters: in both cases the clustering properties result different compared to the second parts of the epochs, but based on Epoch Ia the eruptions produce fewer and smaller clusters with possibly shorter duration, while based on Epoch IIIa the eruptions may constitute fewer, long lasting, overlapping e slightly larger clusters.

FIRST PARTS OF ERUPTIVE EPOCHS

Statistics/Eruption record	Epoch Ia			Epoch IIIa			Epochs Ia x IIIa			Epochs Ia x II x IIIa		
	5th %ile	mean	95th %ile	5th %ile	mean	95th %ile	5th %ile	mean	95th %ile	5th %ile	mean	95th %ile
1/base rate (1/λ0) [a]	97	171	305	102	142	195	101	145	212	92	124	176
self excitement duration (T) [a]	15	333	875	145	970	1385	105	856	1445	38	452	919
mean offspring (μ)	0.08	0.22	0.32	0.31	0.49	0.57	0.22	0.49	0.69	0.15	0.38	0.61
Probability of MN offspring after 477 years	0.0%	1.2%	4.2%	0.0%	7.5%	15.5%	0.0%	7.6%	18.0%	0.0%	2.5%	8.5%
Probability of having 477 years with no events after MN	0.5%	5.4%	16.6%	0.7%	2.5%	5.9%	0.6%	2.8%	7.2%	0.4%	1.8%	4.7%

Table 4.3. Maximizing the likelihood of the first parts of Epoch I (first column) or Epoch III records (second column), or also maximizing the product of the likelihoods of such records together with Epoch II (third column), or even excluding it (fourth column), we report some results. We include: the mean holding time of the new clusters $1/\lambda_0$, the duration of the self-excitement T, the mean offspring of each event μ (first three rows); the probability Q_{mn} for an eruption of producing the first offspring event after 477 years (fourth row); the probability L_{mn} of passing 477 years after an eruption without observing other events anywhere in the caldera (fifth row). Mean values, 5th and 95th percentiles are reported, as a function of the sources of epistemic uncertainty considered.

From Epoch Ia, *i.e.* the first 11-13 eruptions of Epoch I, it is obtained a decrease of the base rate to one on 171 years, [-45%, +80%], of the duration $T = 333$, [-95%, +165%], and of the parameter $\mu = 0.22$, [−65%, +45%], respect to the values calculated on the whole epoch; this implies the decrease of Q_{mn} to [0.0%, 1.2%, 4.2%] and the increase of L_{mn} to [0.5%, 5.4%, 16.6%]. The Epoch IIIa eruptive record, *i.e.* the first 10-11 eruptions of Epoch III, presented a base rate that is 2/3 of the base rate obtained considering the whole epoch: one on 142 years, [-30%, +35%], but a very long duration T of 970 years with an uncertainty range of [-85%,+45%], and a slightly increased $\mu = 0.49$, [-35%, +15%]. The probability Q_{mn} increases to [0.0%, 7.5%, 15.5%] due to the much longer duration of self-excitement, while likelihood L_{mn} is still low at [0.7%, 2.5%, 5.9%].

Maximizing the combined likelihood of Epoch Ia and Epoch IIIa records produces results consistent with Epoch IIIa, but more uncertain (similarly to Epoch Ia): it is obtained a quite low base rate of one on 145 years, [-30%, +45%], a long uncertain duration T of 956 years, [-90%, +70%], and a quite large μ of 0.49, [-55%, +35%]; this corresponds to a decreased number of overlapping quite big clusters. In this case L_{mn} rises to [0.6%, 2.8%, 7.2%], and Q_{mn} to [0.0%, 7.6%, 18.0%]. Maximizing the combined likelihood also including Epoch II record reduces $1/\lambda_0$, μ, and halves the duration T; both Q_{mn} and L_{mn} decrease in average to 2.5% and 1.8%, respectively.

The western and eastern sectors. In Table 4.4 are shown also the parameters obtained considering separately the western and the eastern parts of the Campi Flegrei caldera (zones 1-5 and 6-13 accordingly to the partition in Figure 4.1). During Epoch IW, *i.e.* the western sector activity during Epoch I (8 events), the activity presents a lower μ than in the rest of the Epoch 0.18, [-65%, +65%], but a much shorter and very uncertain duration $T = 101$ years, [-99%, +195%], and a low base rate of one on 378 years, [-30%, +50%]; the clustering behaviour is again not strongly recognizable. The probability Q_{mn} become negligible. The reduction of the base rate increases very much the likelihood L_{mn} to [14.1%, 24.1%, 36.4%] for Epoch IW. Focusing on Epoch IE, *i.e.* the record of eastern sector events, shows a 3/5 base rate and a 1.5 times higher mean offspring than considering the whole Epoch I, and a double duration $T = 1282$ years, [-70%, +45%]; it is likely the presence of some overlapping clusters of events in different zones.

(a)   WESTERN SECTOR

Statistics/Eruption record	Epoch IW			Epoch IIIW			Epochs IW x IIIW			Epochs IW x IIW x IIIW		
	5th %ile	mean	95th %ile	5th %ile	mean	95th %ile	5th %ile	mean	95th %ile	5th %ile	mean	95th %ile
1/base rate (1/λ0) [a]	273	378	559	357	467	698	360	469	709	0	468	697
self excitement duration (T) [a]	0	101	300	4140	5722	7843	0	36	150	0	68	175
mean offspring (μ)	0.06	0.18	0.28	0.07	0.15	0.22	0.00	0.08	0.20	0.00	0.14	0.23
Probability of MN offspring after 477 years	0.0%	0.0%	0.1%	5.3%	10.5%	15.6%	0.0%	0.0%	0.0%	0.0%	0.0%	0.0%
Probability of having 477 years with no events after MN	11.4%	14.1%	36.4%	25.4%	35.9%	48.8%	23.8%	33.8%	46.3%	22.6%	32.8%	45.5%

(b)   EASTERN SECTOR

Statistics/Eruption record	Epoch IE			Epoch IIIE			Epochs IE x IIIE			Epochs IE x IIE x IIIE		
	5th %ile	mean	95th %ile	5th %ile	mean	95th %ile	5th %ile	mean	95th %ile	5th %ile	mean	95th %ile
1/base rate (1/λ0) [a]	151	246	407	75	106	156	100	150	231	99	140	207
self excitement duration (T) [a]	410	1282	1890	8	91	188	50	366	805	62	352	755
mean offspring (μ)	0.31	0.47	0.58	0.31	0.46	0.56	0.39	0.62	0.79	0.38	0.59	0.76

Table 4.4. Maximizing separately the likelihood of the parts of Epoch I (first column) or Epoch III (second column) records located in the (a) western sector and (b) the eastern sector, or also maximizing the product of the likelihoods of such records together respectively with the western and eastern parts of Epoch II record (third column), or even excluding it (fourth column), we report some results. In (a) we include: the mean holding time of the new clusters $1/\lambda_0$, the duration of the self-excitement T, the mean offspring of each event μ (first three rows); the probability Q_{mn} for an eruption of producing the first offspring event after 477 years (forth row); the probability L_{mn} of passing 477 years after an eruption without observing other events anywhere in the caldera (fifth row). In (b) we include the mean holding time of the new clusters $1/\lambda_0$, the duration of the self-excitement T and the mean offspring of each event μ. Mean values, 5th and 95th percentiles are reported, as a function of the sources of epistemic uncertainty considered.

Epoch IIIW is constituted by only three events (Capo Miseno, Averno 1, Averno 2) and from these few eruptions it is not possible to robustly recognize a clustering behaviour: μ decreases to 0.15 with uncertainty range [-55%, +45%] and the duration T becomes huge: 5722 years in average, meaning that in principle the whole epoch could coincide with the cluster length; the base rate is only one on 467 years, [-25%, +50%]. The probability Q_{mn} increases to [5.3%, 10.5%, 15.5%] for Epoch IIIW because of the very long cluster duration, and the likelihood L_{mn} reaches values of [25.4%, 35.9%, 48.8%]. The eastern events produce instead very similar parameters to the whole epoch: the elimination of only three events from the record seems not to change the system behaviour.

Maximizing the combined likelihoods of Epoch IW and IIIW confirms a not recognizable clustering, and a low base rate of one on 469 years, [-25%, +50%], similar to Epoch IIIW; considering also the Epoch IIW record (only 2 events) does not change remarkably the estimates. The probability Q_{mn} is negligible like from Epoch IW record, while the estimates of L_{mn} are much similar to the ones of Epoch IIIW. Epoch IE and IIIE together produce a very high μ of 0.62, [-35%, +25%] with a quite long duration of 366 years, [-85%, +120%], and a base rate of one on 150 years, [-35%, +55%]. Considering also Epoch IIE does not change the results.

Rejecting the epochs hypothesis. Moreover a very strong assumption about the eruption record is the separation in three eruptive epochs. It is remarkable that the Monte Nuovo event occurred such a long time after the end of Epoch III that rejecting its separation from that epoch implies the rebuttal of the separation of the other epochs, as was obtained by the exploration of data. In all the previous cases the ML parameters were fitted on the single epochs or otherwise on more epochs assumed as separate independent samples, *i.e.* multiplying the likelihoods. In contrast in Table 4.5 are reported the results obtained assuming to merge more epochs (and even Monte Nuovo eruption) in a unique record that takes into account also the duration of the periods of quiescence; this corresponds to the idea of rejecting the subdivision between Epoch I and Epoch II, or also all of them.

Focusing on Epoch I*II, *i.e.* all the eruption record between the start of Epoch I and the end of Epoch II, the base rate and the mean offspring are similar to Epoch I values (both slightly lower), and duration T decreases to 359 years, [-85%, +90%]; a possible motivation is the necessity of increasing the likelihood of the quite long interruption of the events during the period of quiescence. Considering the combined likelihood of Epochs (I*II)×III, which corresponds to assume the existence of only two epochs of activity instead of three, gives parameters not much different from the estimates coming from Epoch I×II×III or I×III, with a light decrease of the base rate and a light increase of the mean offspring and the duration.

COMPLETE RECORD, PERIODS OF QUIESCENCE INCLUDED

Statistics/Eruption record	Epoch I*II			Epochs (I*II) x III			Epoch I*II*III			Epoch I*II*III*MN		
	5th %ile	mean	95th %ile	5th %ile	mean	95th %ile	5th %ile	mean	95th %ile	5th %ile	mean	95th %ile
1/base rate (1/λ0) [a]	124	163	227	95	121	159	216	260	314	303	352	404
self excitement duration (T) [a]	53	359	685	65	215	478	103	292	475	152	337	480
mean offspring (μ)	0.15	0.28	0.38	0.34	0.49	0.65	0.33	0.42	0.49	0.38	0.45	0.51
Probability of MN offspring after 477 years	0.0%	0.9%	3.3%	0.0%	0.6%	2.4%	0.0%	0.6%	2.0%	0.0%	0.8%	2.3%
Probability of having 477 years with no events after MN	1.6%	4.6%	9.5%	0.4%	1.4%	3.1%	7.2%	10.6%	14.7%	13.2%	16.4%	19.8%

Table 4.5. Maximizing the likelihood of the merged records of Epoch I and Epoch II including the period of quiescence duration (first column), maximizing the product of the likelihood of this merged record with Epoch III (second column), maximizing the likelihood of the merged records of all the epochs (third column), or even including Monte Nuovo event still including the periods of quiescence (fourth column), we report some results. We include: the mean holding time of the new clusters $1/\lambda_0$, the duration of the self-excitement T, the mean offspring of each event μ (first three rows); the probability Q_{mn} for an eruption of producing the first offspring event after 477 years (fourth row); the probability L_{mn} of passing 477 years after an eruption without observing other events anywhere in the caldera (fifth row). Mean values, 5th and 95th percentiles are reported, as a function of the sources of epistemic uncertainty considered.

Based on Epoch I*II*III record, which corresponds to the extreme assumption of refusing every subdvision in epochs, the results change a lot: the base rate is halved, and considering the complete record that includes even a third period of quiescence and Monte Nuovo eruption, the base rate is only one on 352 years, [-15%, +15%], with duration 337 years, [-55%, +40%], and mean offspring 0.45, [-15%, +10%]. The probability Q_{mn} is [0.0%, 0.8%, 2.3%], but these parameters increase the likelihood L_{mn} to [13.2%, 16.4%, 19.8%]. Anyways the non-homogeneity between the eruptive epochs and the periods of quiescence seems too strong to be captured by such a simple model: the rejecting of the epochs hypothesis would require additional mathematical features aimed at increasing the likelihood of such long interruptions of the activity.

4.5.2. Probability distribution of next eruption time

The results about next eruption time forecast are a very complex issue and strongly rely on the implications that the volcanological assumptions have on the Cox-Hawkes process parameters. However, based on the model developed as reconstructed on the available datasets, it is possible to estimate the remaining time before the next Campi Flegrei eruption. In particular, the estimates include the residual self-excitement coming from Monte Nuovo eruption: excitement coming from previous eruptions is assumed negligible because of the long period of quiescence. The doubly stochastic structure of the model permits to separate the epistemic uncertainty from the physical variability: like in the previous chapters, we represented the physical variability of an observable, for which we produced percentile estimates that quantified the sensitivity of the results on the main sources of epistemic uncertainty.

All the curves reported in Figure 4.8 correspond to ML exponential distributions, but alternative non-parametric results coming from gaussian kernel density estimators are completely consistent with the exponential curves. In Figure 4.8a the sensitivity to different assumptions concerning the time windows of the eruption records on which the likelihood is maximized are reported. Considering the three eruptive epochs as independent samples and assuming Monte Nuovo as the first event of a new epoch of activity, we obtain a mean time to the next Campi Flegrei eruption of 103 years, with physical variability ranging from 5 to 318 years; the epistemic uncertainty have been quantified as [-25%, +35%] of these values (black lines). As expected from the results on the parameters of the process, focusing on the first part of the epochs produces a slightly slower eruption rate (red lines), whereas including the periods of quiescence produces waiting times 3 times longer (purple lines).

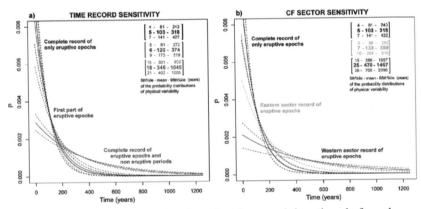

Figure 4.8. Probability distributions for the remaining time before the next eruption, assuming ML exponential distributions on the time samples of the described model. The bold lines indicate the mean probability density functions per year, and the dashed lines are composed of the 5th and 95th epistemic uncertainty percentiles of the values of such functions. Different colours correspond to alternative volcanological assumptions. In (a) are compared different time records, in (b) are compared eastern and western sectors records. The values reported are the 5th percentile, the mean and the 95th percentile with respect to epistemic uncertainty (from above to below), of the 5th percentile, the mean value and the 95th percentile of the physical variability (from left to right).

In Figure 4.8b is reported instead the sensitivity of the time forecast by considering a separation between the eastern and western portions of the caldera. Considering a probability model with different parameters for the eastern and western sectors produces estimates which are very similar to the ones from the model that does not separate the record. Focusing only on the eastern sector dataset tends to slow a bit the eruption rate (green lines), and the western record instead rises the time estimates of almost five times to 470 years in mean, with physical variability ranging from 25 to 1467 years; epistemic uncertainty have been quantified as [-25%, +45%] of these values (blue lines).

Alternatively assuming to start the simulation immediately after Monte Nuovo event and then rejecting the samples corresponding to times before the present, produces time estimates that are increased of +15% concerning the complete record of the eruptive epochs (119 years in mean), but only of +5% in the case of the western sector record (508 years in mean). Indeed the epistemic assumptions corresponding to shorter holding times are rejected more easily. However this approach no more relies directly on the doubly stochastic probability model that we defined based on past eruptive record.

4.6. Appendix A: Formal definition of the uncertainty model

The model for epistemic uncertainty affecting past eruptions times record is obtained by a simple Monte Carlo simulation based on several conditional samplings. In the sequel we formally report the technical details of such model. First we define 3 classes of eruptions with different levels of information, which will be treated separately.

Definition 4.10 (Classes of events). Class $C_1 = (w_i)_{i=1,...,n_1}$ includes n_1 events possessing datation and ordering; class $C_2 = (w_i)_{i=n_1+1,...,n_1+n_2}$ contains n_2 events possessing only the ordering (including some small uncertainties); class $C_3 = (w_i)_{i=n_1+n_2+1,...,n1+n2+n3}$ is formed by n_3 events that possess some datation bounds, but not any robust ordering information. Lets call $n = n_1 + n_2 + n_3$ the total number of events and $m_{12} = n_1 + n_2$ the number of events belonging to the union of the first two classes.

We start the definition of the probability model sampling the ordering of past events of classes C_1 and C_2.

Definition 4.11 (The ordering sampling). We define a random variable $\check{\tau}$ from $(E, \mathcal{E}, \eta)$ on the space $\mathcal{S}(m_{12})$ of the permutations of $\{1, \ldots, m_{12}\}$ such that $(\check{v}_j)_{j=1,...,m_{12}}$, where

$$\check{v}_j := w_{\check{\tau}(j)}, \quad \forall j,$$

represents a random sample for the ordered family of eruptive events, not including class C_3.

The permutation $\check{\tau}(\omega)$ will be for almost every $\omega \in \Omega$ the restriction of the permutation $\tau(\omega)$ of Definition 4.1 on $\mathcal{C}_1 \cup \mathcal{C}_2$. First we sample the times for the events in classes $\mathcal{C}_1$ and $\mathcal{C}_3$.

Definition 4.12 (Events with datation bounds). Let $([a_i, b_i])_{i=1,\dots,n_1}$ and $([a_i, b_i])_{i=m_{12}+1,\dots,n}$ be closed intervals on $\mathbb{R}$ representing the datation windows available for the elements of $\mathcal{C}_1$ and $\mathcal{C}_3$, each associated with a central value t_i^* (the midpoint between a_i and b_i in absence of other information). Let $(s_i)_{i=1,\dots,n_1}$ and $(s_i)_{i=m_{12}+1,\dots,n}$ be vectors of real random variables on $(E, \mathcal{E}, \eta)$, each s_i representing the time of eruptive event w_i. We assume that each s_i has a triangular distribution supported on the interval $[a_i, b_i]$ with central value t_i^*, and that $s_{i_1} < s_{i_2} \; \forall i_1, i_2 \leq n_1$ such that $\check{\tau}^{-1}(i_1) < \check{\tau}^{-1}(i_2)$, *i.e.* the time samples are consistent with the ordering samples.

In the following we complete the time sampling with the elements of $\mathcal{C}_2$, which do not possess any temporal bounds except for their ordering.

Definition 4.13 (Events without datation bounds). Let (w_{i_1}, w_{i_2}) be a pair of events in class $\mathcal{C}_1$, such that $v_{\check{\tau}^{-1}(i_1)} < v_{\check{\tau}^{-1}(i_2)}$ and they are adjacent in $\mathcal{C}_1$, *i.e.* does not exist a separator inside $\mathcal{C}_1$ with respect to their ordering. Let $W = (w_i)_{i \in I}$ where $I := \check{\tau}\{k | \check{\tau}^{-1}(i_1) < j < \check{\tau}^{-1}(i_2)\}$ be the set of events in class $\mathcal{C}_2$ which are ordered between w_{i_1} and w_{i_2}. Then we define the random vector $(s_i)_{i \in I}$ as a number of independent random variables uniformly distributed on the interval $[s_{i_1}, s_{i_2}]$, ordered increasingly. We repeat this procedure for each adjacent pair in class $\mathcal{C}_1$. For defining even the times of events of class $\mathcal{C}_2$ ordered before the first or after the last event of class $\mathcal{C}_1$, we assumed a time interval uniformly distributed on $[0, 100]$ years between each of them.

We got the time samples $(s_i)_{i=1,\dots,n}$ for all the events; the last phase of time sampling completes the procedure, re-ordering them.

Definition 4.14 (The joint distribution of times). We extend the permutation $\check{\tau}$ on the whole set of n events, including the events of class $\mathcal{C}_3$ consistently with the order of the sampled times, and we call it τ. Then we define

$$v_j := w_{\tau(j)}, \quad t_j := s_{\tau(j)}, \forall j = 1, \dots n,$$

as the randomly ordered events with their random times.

Moreover we report that for each $j \leq n$ the distribution for the spatial location V_j of Definition 4.1 is the uniform probability measure ζ_j on the set D_j of Definition 2.6 (see also Figure 4.1). We remark that this procedure is the core of every Monte Carlo simulation assessed in this chapter.

Chapter 5
Addendum

5.1. Summary

In this Chapter are included three detailed digressions about important topics concerning the mathematical and physical background of the previous chapters. In particular is presented a general introduction to the expert elicitation theory, then the complete construction of the simplified model adopted for propagating pyroclastic density currents (box model), and at last a summary about Cox processes and Hawkes processes.

5.2. The expert judgment approach

By the term expert elicitation we mean those techniques used to inform decisions, forecasts or predictions based on a formalized treatment of the judgments or opinions of experts, usually in the context of decision support for non-specialists (*e.g.* [41, 92, 120, 8]). Performance based elicitation procedures include an empirical step of expert ranking: based on the answers given by the experts to test questions with known answers (called seed items), different weights are computed and attributed to individual experts (*e.g.* [42, 43]). These weights are then used to pool their opinions about target questions of specific interest to a problem owner, *e.g.* those that cannot be determined from data or direct measurement, or those that will become known only in the future, after a decision has been made. In practice, the goals of an elicitation can be twofold: to give reliable pointwise estimates of variables of interest and to assess the level of uncertainty of such estimates. Here, we mean a sort of intrinsic uncertainty, not the subjective precision of the estimate declared by an expert who, by character, may be overconfident, proficient, or cautious in making judgments (see also [95]).

Some experts may be very good in estimating true values, while others may be better in the assessment of the true uncertainty; in the first case we say they are good in accuracy, in the second we say they are well calibrated. When we consider the performance of a particular weighting method on the test questions, we would like to measure both accuracy and calibration; but these features cannot be summarized in a single index: they are inherently bi-dimensional (see [68]). These two abilities have a different nature and may pertain to different people: for instance, very good point value estimators may be overconfident when it comes to uncertainty estimation. For this reason, it is possible to measure the performance of a weighting method by means of various indices that reflect different features: we define three indexes called calibration, informativeness, and expected accuracy, the meaning of each reward will be explained in detail in the sequel.

Among the several weighting methods known in the literature, we focus on the Cooke Classical model well described in [41], because it is devised on the basis of formal performance scoring rules, incorporates empirical control and is found on the calibration index. We use also the Expected Relative Frequency (ERF) method of [68], which in some sense is based on likelihood. It is based on an orthodox probabilistic idea related to probability densities: from the answers of the expert we build up a probability density function and then is possible to compute the likelihood of the true answer under this density. However, the likelihood-based methods can have a practical drawback: when the density is very concentrated, very peaked, its values are very large (much greater than one) and the weight of the expert may be implausibly high; and it can be criticized theoretically, since one could argue that an expert inclined to accept high risks is encouraged to give very concentrated densities to receive huge rewards. For these reasons the density is integrated on a small interval around the true value to diminish these problems. This is the number computed by the method as the Expected Accuracy index.

Still in [68] are observed the following tendencies: i) group-based models (the Cooke Classical model, ERF model and also Equal Weights) perform better than single Best Expert models and the majority of single a priori-chosen experts; ii) non-trivial models are the best, in most situations, with respect to specific aims of elicitation (Cooke Classical model for the assessment of uncertainty, ERF model for pointwise estimation); iii) the performance of the Equal Weights model is often close to the optimal ones. For these last reason we always report also the results obtained with this basic approach.

5.2.1. Mathematical preliminaries

An elicitation session involves a group $\mathcal{E}$ of k experts, who are asked to answer a questionnaire $\mathcal{Q}$. The questionnaire is composed of two type of questions: a set $\mathcal{Q}_{sd}$ of seed questions and a set $\mathcal{Q}_{trg}$ of target questions. The number n above is the cardinality of $\mathcal{Q}_{sd}$. Seed questions have known answer (known to the interviewer) and have the purpose of giving weights to the experts. The target questions are those of interest in the specific field of application, with unknown answer. The numbers $(x_q)_{q=1,...,n}$ above are the true answers to the seed questions. Each expert e is uncertain about each question q. Uncertainty may have a physical origin, due to the complexity of phenomena behind the problem, or may be just subjective, due to partial knowledge of the problem, of some of its input data. All this uncertainty is summarized in the pdf $f_{e,q}$, which depends on expert and question. So $f_{e,q}$ is the subjective statistical evaluation of question q from expert e.

Remark 5.1. We have described the answers of experts by pdfs. This is what experts are asked to tell us in their questionnaires: summaries of $f_{e,q}$. Ideally, however, we may also think that experts are asked to give a numerical answer, a guess of the value x_q itself, not its statistics. Let us call $X_{e,q}$ the random answers we expect to receive from expert e for question q. The law of $X_{e,q}$ is represented by $f_{e,q}$, by definition of these two objects. The random variable $X_{e,q}$ has an auxiliary character in expert elicitation, since it is never really asked, but it is real in potential applications of this theory to pooling the results of numerical simulations from different codes ($X_{e,q}$ is the sample value, $f_{e,q}$ is only estimated a posteriori if we have a large number of samples).

Denote by L_+^1 the set of all pdfs on real numbers, namely all Borel measurable functions $f : \mathbb{R} \to [0, +\infty)$ such that $\int_{-\infty}^{\infty} f(x)\,dx = 1$. Denote by $\mathcal{P}_k$ the set of all vectors $(w_e)_{e=1,...,k} \in [0, +\infty)^k$ such that $w_e \geq 0$ and $\sum_{e=1}^{k} w_e = 1$.

Definition 5.2 (Reward). Given $n \in \mathbb{N}$, any function

$$F : \left(L_+^1\right)^n \times \mathbb{R}^n \to [0, +\infty)$$

will be called a reward on n questions (we accept also the case when the domain of F is just a subset of $\left(L_+^1\right)^n \times \mathbb{R}^n$).

A reward is the score we give to an expert who chose certain pdfs to describe the answers to the seed questions. The score is computed by a comparison between the pdf $\left(f_{e,q}\right)_{q=1,...,n}$ and the true answers $(x_q)_{q=1,...,n}$; thus, to compute the score we use n densities and n real numbers. Of course the definition or reward is 'empty' in a sense, since any map $F : \left(L_+^1\right)^n \times \mathbb{R}^n \to [0, +\infty)$ (even far from the idea of reward) has this name. The aim of the theory is to find out useful rewards. The weighting rule associated to a reward is an obvious concept: if the reward gives scores $W_1, \ldots, W_k$ to the k experts, we just normalize them to define weights (weights are non-negative numbers with sum one).

Definition 5.3 (Weighting rule). Let F be a reward on n questions and $k \in \mathbb{N}$, the corresponding weighting rule is the function

$$W_F : \left(L_+^1\right)^{n \times k} \times \mathbb{R}^n \to \mathcal{P}_k$$

defined as

$$W_F \left(\left(f_{e,q}\right)_{e,q}, \left(x_q\right)_q\right) = \frac{F\left(\left(f_{e,q}\right)_q, \left(x_q\right)_q\right)}{\sum_{e'} F\left(\left(f_{e',q}\right)_q, \left(x_q\right)_q\right)}.$$

where we have shortened the notations

$$\left(f_{e,q}\right)_{e,q} = \left(f_{e,q}\right)_{\substack{e=1,\dots,k \\ q=1,\dots,n}}, \quad \left(f_{e,q}\right)_q = \left(f_{e,q}\right)_{q=1,\dots,n}, \quad \left(x_q\right)_q = \left(x_q\right)_{q=1,\dots,n}$$

Whence weights

$$\left(w_e\right)_{e=1,\dots,k} = W_F \left(\left(f_{e,q}\right)_{e,q}, \left(x_q\right)_q\right)$$

have been found, consider any one of the target questions (the argument applies to each of them). The experts have chosen densities $\left(f_e\right)_{e=1,\dots,k}$ for the answer to it. A pooling rule is a way to average these densities and get a new one, say f.

Definition 5.4 (Pooling rule). Given $k \in \mathbb{N}$, any function

$$G : \left(L_+^1\right)^k \times \mathcal{P}_k \to L_+^1$$

will be called a pooling rule on k experts.

The result $G\left(\left(f_e\right)_{e=1,\dots,k}, W_F\left(\left(f_{e,q}\right)_{e,q}, \left(x_q\right)_q\right)\right)$ of this series of operations has been called forecast map, since f is the density which describes the unknown answer to that question. The forecast map requires the choice of two basic ingredients: a reward and a pooling rule. We call Decision Maker any such choice.

Definition 5.5 (Decision Maker). Given $n, k \in \mathbb{N}$, a Decision Maker (DM) is any pair (F, G) where F is a reward on n questions and G a pooling rule on k experts. The DM (F, G) defines a map

$$\Phi_{F,G} : \left(L_+^1\right)^{n \times k} \times \mathbb{R}^n \times \left(L_+^1\right)^k \to L_+^1$$

given by

$$\Phi_{F,G}\left(\left(f_{e,q}\right)_{e,q}, \left(x_q\right)_q, \left(f_e\right)_{e=1,\dots,k}\right)$$
$$= G\left(\left(f_e\right)_{e=1,\dots,k}, W_F\left(\left(f_{e,q}\right)_{e,q}, \left(x_q\right)_q\right)\right).$$

We call it the forecast map.

Often in practice we do not assume that the expert knows $f_{e,q}$ in detail. We only assume expert e is able to declare certain summaries of $f_{e,q}$, like the 5%, 50% and 95% percentiles. Therefore, while in this preliminary part we work prevalently with the full pdf $f_{e,q}$, when we will focus on applications and real data we use approximations based on summaries of $f_{e,q}$ (experts are not able to quantify their beliefs with a full pdf).

Definition 5.6 (Representation map). Given $m \in \mathbb{N}$, and a sequence $(t^1 < \cdots < t^m) \in [0, 1]^m$, any symmetrical function (*i.e.* not depending of the order of components of the variable vector)

$$R : \mathbb{R}^m \to \left(L_+^1\right)$$

will be called a representation map on the quantiles $(t^i)_{1 \le i \le m}$ if

$$\int_{-\infty}^{v_i} [R(v)](u)du = t^i,$$

for each $i \in \{1, \ldots, m\}$.

All these definitions are quite generic, as it is the definition of estimator in statistics. A reward may be good for some purposes and less for others; the order between DMs may depend on the purpose. We have detected experimentally a dichotomy between good DMs for pointwise estimate and good ones for uncertainty assessment. Therefore, it seems it is not possible to give such general concepts in a unique way.

5.2.2. Equal weights rule

We start with this simple method, that essentially combines the experts opinions giving equal weight to them.

Definition 5.7 (Equal reward). Let F be a reward on n questions, we call it equal reward if it is constant. The constant value of F does not matter, since we always have

$$W_F \left(\left(f_{e,q}\right)_{e,q}, \left(x_q\right)_q \right) = \frac{1}{k}.$$

The pooling rule that is used is very natural, it is simply the mixture of the given densities.

Definition 5.8 (Linear pooling). Define the pooling rule

$$G_l \left((f_e)_{e=1,\ldots,k}, (w_e)_{e=1,\ldots,k} \right) = \sum_{e=1}^{k} w_e f_e.$$

This is called linear pooling.

Linear pooling is the standard choice in part of the literature, see for instance [41]. It is correct that linear pooling is good for incorporating both the uncertainty of each expert and the variability between them (suitably weighted). Indeed on some probability space (Ω, F, P), let $(X_e)_{e=1,\dots,k}$ be independent random variables on such that X_e has law f_e, and I be a random variable taking values in $\{1, \dots, k\}$ with $P\,(I = e) = w_e$. Then the random variable X_I has law $\sum_{e=1}^{k} w_e f_e$. In other words, $\sum_{e=1}^{k} w_e f_e$ is the law of a mixture. With probability $w_{e'}$ we choose the answer of expert e', with probability $w_{e''}$ the answer of expert e'', and so on.

Remark 5.9. In Definition 5.22 is also shown the quantile pooling, an alternative rule that focuses on the accuracy of pointwise estimation losing an amount of the spread of the ranges of uncertainty. A third drastic approach not followed by any of the method presented may be to average the answers, $\sum_{e=1}^{k} w_e X_e$, to take advantage of an averaging property similar to the Law of Large Numbers. This could be called 'weighted convolution pooling rule', but has the big drawback of strongly diminishing the uncertainty range of the associated DM with respect to the ranges of the single experts; in the trivial case of all the experts giving for each question independent identically distributed answers this DM would have a small spreading around the mean, smaller than the range of the single experts and tending to zero as the number k of the experts grows.

At this point we can take the Decision Maker associated to the previous definitions.

Definition 5.10 (Equal weight DM). We call Equal Weight DM the pair (F, G) where F is constant and G is linear pooling. Therefore

$$\Phi_{F,G}\left((f_{e,q})_{e,q}, (x_q)_q, (f_e)_{e=1,\dots,k}\right) = \frac{1}{k}\sum_{e=1}^{k} f_e.$$

Associated to this rule there is a simple representation map.

Definition 5.11 (Maximum entropy distribution). Given $m \in \mathbb{N}$, a sequence $(t_1 < \cdots < t_m) \in [0, 1]^m$ and a range $[a, b]$, we define the representation map $R(v)$ as the piecewise constant function that on each interval $(v_{i-1}, v_i]$, $1 \le i \le m + 1$ is equal to $\frac{t_i}{v_i - v_{i-1}}$, where $(v_0, v_{m+1}) = (a, b)$. It is trivial that it is a representation map on the fixed quantiles, and it is called maximum entropy distribution (or minimal information distribution).

It will be observed that with this representation map each variable must be supplied with an intrinsic range $[a, b]$ containing all the quantiles elicited from experts. In some cases the choice of the cutoff points might be motivated, for example if the answer $X_{e,q}$ were a relative frequency or a percentage then 0 and 1 might be natural cutoff points. However, in some cases the choice of cutoff points must be made ad hoc. In our case when $X_{e,q}$ does not represent percentages we used a 10% overshoot above and below the interval $[v_1, v_m]$. See more details of such representation map on [41].

5.2.3. Cooke Classical Model

The Cooke Classical model is based on a reward that is the combination of the different indexes called calibration and informativeness, with calibration playing a dominant role. The designation classical derives from a close relation between calibration scoring and hypotheses testing in classical statistics. Initial development was sponsored by the Dutch Ministry for environment, and later phases have been supported by the European Space Agency and the European Community. Now we introduce the concept of relative information, needed for the definitions of both calibration and informativeness. Let $p = p_1, \ldots, p_n$ be a probability distribution over alternatives $\{1, \ldots, n\}$ the entropy or negative information $H(p)$ of p is defined as

$$H(p) := -\sum_{i=1}^{n} p_i \ln p_i.$$

Let $s = s_1, \ldots, s_n$ be a probability distribution, and assume $p_i > 0$ for all i, then the relative information of s with respect to p is defined as

$$I(s, p) := \sum_{i=1}^{n} s_i \ln \frac{s_i}{p_i},$$

and we have that $I \geq 0$ and it is null is and only if $s = p$. It is commonly taken as an index of the information learned if one initially believes that p is correct and subsequently learns that s is correct. It is also true that $H(p) = \ln n - I(p, u)$ where u is the uniform distribution over $1, \ldots, n$; that is $u_i = 1/n$ for each i. Similarly, if f_1 and f_2 are pdfs and $f_2 > 0$ on the Borel set $A \subseteq \mathbb{R}$ the relative information of f_1 respect to f_2 is defined by

$$I(s, p) := \int_A f_1(u) \ln \frac{f_1(u)}{f_2(u)} du.$$

This method is strongly based on applications, and the following construction assumes that each expert e give a vector of $m \in \mathbb{N}$ values $(f_{e,q}^i)_{1 \leq i \leq m}$ for each question q, associated to the quantiles $(t_i)_{1 \leq i \leq m}$.

The representation map R adopted is the maximum entropy distribution defined above. Define $p_i = t_i - t_{i-1}$, $i = 1, \ldots, m+1$, where $(t_0, t_{m+1}) := (0, 1)$. Assume $p = (p_1, \ldots, p_{m+1})$, and call probability outcome each interval $[t_{i-1}, t_i]$; the vector p represents a probability measure on the probability outcomes.

Definition 5.12 (Informativeness). The informativeness index of a function f on a fixed range $[a, b]$ is the relative information with respect the uniform distribution on $[a, b]$. If $f = R(v)$, we have that

$$I(R(v))_{[a,b]} = \ln(b-a) + \sum_{i=1}^{m+1} p_i \ln \frac{p_i}{v_i - v_{i-1}},$$

where R is the representation map, p is defined above and v is the quantile values associated. The informativeness index of an expert e with respect to set of questions Q is the average of the informativeness indexes with respect to the questions (all assumed with an intrinsic range):

$$I(e) = \frac{1}{n} \sum_{q=1}^{n} I(f_{e,q}).$$

We now define a general class of rewards, focused on uncertainty estimation.

Definition 5.13 (H-based reward). Let $H : [0, 1]^n \to [0, +\infty)$ be a function which measure an inverse distance between a sample $(u_q)_{q=1,\ldots,n}$ and the uniform distribution. By 'inverse distance' we mean that samples 'closer to uniform ones' have larger values of H. Then consider the reward

$$F\left((f_q)_{q=1,\ldots,n}, (x_q)_{q=1,\ldots,n}\right) = H\left(\left(\int_{-\infty}^{x_q} f_q(x)\, dx\right)_{q=1,\ldots,n}\right).$$

We call it H-based uncertainty estimate reward.

This is the basic idea to reward well those experts who capture the correct level of uncertainty of a problem. Let us briefly describe a general idea to give a meaningful score to the ability in uncertainty assessment, formalized above with the H-based uncertainty estimate reward. We have to think that the true answers x_q are realizations of random variables X_q and the best possible expert is the one who captures their probability distribution. Then, in the best possible case, $f_{e,q}$ should be the pdf of X_q. If so, the pdf of the random variable $\int_{-\infty}^{X_q} f_{e,q}(x)\, dx$ is uniform on $[0, 1]$, by a well known theorem. This motivates the following idea: let e be an expert, and for each seed question $q \in \mathcal{Q}_{sd}$, compute the cumulative probabilities corresponding to the true answers

$$p_{e,q} = \int_{-\infty}^{x_q} f_{e,q}(x)\, dx.$$

As we said above, in the best possible case the set of numbers $(p_{e,q})_{q=1,\dots,n}$ should be a random sample from the uniform distribution on $[0, 1]$. Hence, reward rules suitable to score ability in uncertainty assessment may be based on a numerical measure of the anomaly of the set of numbers $(p_{e,q})_{q=1,\dots,n}$ with respect to uniform distribution (anomalous sets should correspond to low score). The calibration index heuristically belongs to this class of rewards, however it is not a function of a single pdf (the answer of the expert) and a value (the true answer), but rather it is a score of the single expert depending of all the seed questions; if normalized it would be a weighting rule. We remind the idea of the Chi-square test, contained in the concept of calibration index. Let s denote a sample distribution generated by n independent samples from the distribution p. Let χ_d^2 denote the cumulative distribution function a chi square variable with d of freedom. Then we have that

$$P\{2NI(s, p) \le x\} \to \chi_{n-1}^2(x), \quad N \to +\infty.$$

This property imply a simple test that may be used as an inverse distance, defining n independent samples from the true answers $(x_q)_{q=1,\dots,n}$. As said above, assume that any pdf $f_{q,e} = R((f_{e,q}^i)_{1 \le i \le m})$ for some quantiles $(t_i)_{1 \le i \le m}$. For each fixed expert e any value x_q of a seed question belongs to an interval $[f_{i-1}, f_i]$ for some $1 < i < m$; in this way, observations of $x_1, \dots, x_n$ generate a sample distribution $s = (s_1, \dots, s_{R+1})$ over the probability outcomes for each expert. In general if the densities $f_{q,e}$ do not come from a representation map, if we take $\varphi_{q,e}$ as its cumulative distribution function it is possible to define p as the uniform distribution on $[0, 1]$ and take the values $(\varphi_{q,e}^{-1}(x_q))_{1 \le q \le n}$ as the sample of the distribution s; in this case the probability outcomes are a continuous set, but in the following we will never need this case.

Definition 5.14 (Calibration). The calibration index of an expert e is defined as

$$C(e) := 1 - \chi_R^2 [2NI(s, p)],$$

where $I(\cdot, \cdot)$ is the relative information function and s, p are the distributions on the probability outcomes defined above.

Finally we can define the complicated weighting rule of this method.

Definition 5.15 (Cooke weighting rule). Let $\alpha \in (0, 1)$. Take

$$W'_e = C(e) \times I(e) \times 1_{[\alpha,1)}(C(e)),$$

and define the Cooke weighting rule as

$$W\left(\left(f_{e,q}\right)_{e,q}, \left(x_q\right)_q\right) = \frac{W'_e}{\sum_{e=1}^{k} W'_e}.$$

The constant α eliminates the experts that are poorly calibrated, representing in some sense is the significance level of the statistical test. For each choice of α a different DM^α is defined. It can also be scored like an additional expert, hence is defined $W\left(\left(f_{DM^\alpha,q}\right)_{e,q}, \left(x_q\right)_q\right)$, the weight that a virtual expert would receive when he gives the DM^α distribution and it is scored along the actual experts. Suppose that we add the virtual expert to the set of experts and re-calculate a new DM: it is easy to see that this new DM would be equal to the initial one. It is possible to take the α value that maximizes the DM virtual weight, thereby optimizing that parameter choice and defining the so called Cooke DM.

Definition 5.16 (Cooke DM). With a little abuse of notation the pair (W, G_l) where W is the Cooke weighting rule and G_l is linear pooling is called Cooke DM.

5.2.4. Expected relative frequency model

The following reward gives high score to those experts who choose densities which give high probability to a neighborhood of the true value, as an index of their pointwise accuracy. In this sense it may be considered similar to a likelihood score. It can be considered opposite to the Cooke method because that is based on uncertainty estimation, while this is based on a pointwise estimation (see [68] for more details about this method).

Definition 5.17 (Expected accuracy reward). Given $\beta > 0$, define, for all $\left(x_q\right)_{q=1,\ldots,n} \in \left(\mathbb{R} \setminus \{0\}\right)^n$ the reward

$$F\left(\left(f_q\right)_{q=1,\ldots,n}, \left(x_q\right)_{q=1,\ldots,n}\right) = \frac{1}{n}\sum_{q=1}^{n}\int_{I_q} f_q(x)\,dx$$

where

$$I_q = \left[x_q(1-\beta), x_q(1+\beta)\right].$$

It is called Expected accuracy (or ERF) reward.

The definition of I_q is multiplicative because the size of the interval must be linked to the scale of the problem, an information embodied by the true answer x_q. If are introduced typical length scales we can easily give an additive version of the previous definition. It is reasonable to expect that this reward give prize to experts who are good in pointwise estimation of the true value. Let the random variable $X_{e,q}$ represent an answer (see Remark 5.1). Let us introduce the concept of accurate answer.

Definition 5.18 (Accurate answer). Choose intervals $(I_q)_{q=1,...,n}$ around the true answers, $x_q \in I_q$. We call accurate answer (for expert e about question q) the event

$$X_{e,q} \in I_q.$$

Expert e has opinion $f_{e,q}$ about question q. Based on this opinion, the probability expert e gives an accurate answer is

$$P\left(X_{e,q} \in I_q\right) = \int_{I_q} f_{e,q}(x)\, dx.$$

Definition 5.19 (Expected Relative Frequency). The relative number of accurate answers for expert e on questionnaire $\mathcal{Q}_{sd}$, that we call empirical score of e on $\mathcal{Q}_{sd}$, is the random variable

$$\text{Score}(e, \mathcal{Q}_{sd}) := \frac{1}{n} \sum_{q=1}^{n} 1_{\{X_{e,q} \in I_q\}}$$

where $1_{\{X_{e,q} \in I_q\}}$ is the indicator function of the good answer event $\{X_{e,q} \in I_q\}$. Let us call Expected Relative Frequency of accurate answers, shortly ERF, the average score:

$$ERF(e, \mathcal{Q}_{sd}) := E\left[\text{Score}(e, \mathcal{Q}_{sd})\right].$$

The number ERF is a measure of the ability of expert e to predict values close to the true ones. An expert with very high ERF chooses densities $f_{e,q}$ such that, in the average, realizations of the family $\left(X_{e,q}\right)_{q=1,...,n}$ have several values very close to the true values (closedness depends on the choice of I_q). The number ERF, a priori, should depend on the joint distribution of the random variables $\left(X_{e,q}\right)_{q=1,...,n}$, but linearity of the average implies it depends only on the family of marginals $\left(f_{e,q}\right)_{q=1,...,n}$:

Proposition 5.20 (Cooke). *We have*

$$ERF(e, \mathcal{Q}_{sd}) = \frac{1}{n} \sum_{q=1}^{n} P\left(X_{e,q} \in I_q\right) = \frac{1}{n} \sum_{q=1}^{n} \int_{I_q} f_{e,q}(x)\, dx.$$

Proof. It simply follows from

$$ERF\,(e, \mathcal{Q}_{\mathrm{sd}}) = E\left[\frac{1}{n}\sum_{q=1}^{n}1_{\{X_{e,q}\in I_q\}}\right] = \frac{1}{n}\sum_{q=1}^{n}E\left[1_{\{X_{e,q}\in I_q\}}\right]. \quad \square$$

Due to this proposition we see that ERF coincides with the ERF reward, explaining its name. Lets pass to the definition of an alternative pooling rule that enhances the accuracy of pointwise estimation reducing the spread of the ranges of uncertainty; the idea is conceptually linked to the equivalent representation of the linear pooling in the space of cumulative distribution functions (cdf).

Remark 5.21. Because of linearity of the integral functional, the linear pooling could also be defined using cdf $\varphi(f)$ as

$$\varphi\left(G_l\left((f_e)_{e=1,\dots,k}, (w_e)_{e=1,\dots,k}\right)\right) = \sum_{e=1}^{k} w_e \varphi(f_e).$$

Similarly, it is possible to take the linear combination of quantile functions $\theta(f, \cdot) := \varphi(f, \cdot)^{-1}$.

Definition 5.22 (Quantile pooling). Define a pooling rule such that:

$$\theta\left(G_q\left((f_e)_{e=1,\dots,k}, (w_e)_{e=1,\dots,k}\right)\right) = \sum_{e=1}^{k} w_e \theta(f_e).$$

This is called quantile pooling.

Hence the linear pooling rule corresponds to take the vertical mean of the graphs of the cdf, while this rule corresponds to take an horizontal mean of that graphs. The purpose of the next example is to show in what sense the quantile pooling focuses on the pointwise accuracy of the associated DM. Se also some examples in the sequel.

Example 5.23. Assume to have two experts with the same weight $\frac{1}{2}$, and probability distributions corresponding respectively to a Dirac δ in zero and the uniform distribution on $[-1, 1]$. Then with the linear pooling is obtained a mixture of the two: with probability $\frac{1}{2}$ the DM will extract with the δ, otherwise with the uniform. This is the best way for defining an uncertainty distribution that is the combination of the two, but nevertheless it could show a bad behaviour with respect to the pointwise estimates. Suppose that the first expert is quite accurate (but enormously overconfident) and the true answer x falls near zero but it is not null: then the accuracy of the DM becomes poor as soon as it is calculated on a small enough interval around x. In contrast the adoption of quantile pooling produces a triangular distribution, with vertex in 0 and base on $[-1, 1]$ and in the case of x near zero but not null the expected accuracy increases significantly.

At this point we can take the Decision Maker associated to the previous definitions.

Definition 5.24 (Expected Relative Frequency DM). We call Expected relative frequency DM the pair (F, G_q) where F is ERF reward and G_q is the quantile pooling rule.

In the very common case of eliciting three quantiles which the second is the median (*i.e.* 5th, 50th and 95th percentiles) the associated representation map of this method is not the maximum entropy distribution defined above, but is a map that tends to centralize more the probability measure constructed around its central value; this enhances the pointwise accuracy of the method. See an example in Figure 5.1.

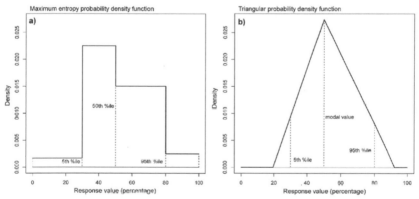

Figure 5.1. Examples of representation maps for an expert response equal to [30% - 50% - 80%]. In (a) is reported the maximum entropy probability distribution, in (b) the triangular probability distribution.

Definition 5.25 (Triangular distribution). Given $m \in \mathbb{N}$, the triplet $(t_1, t_2, t_3) \in [0, 1]^m$, where $t_2 = 0.5$, with a small abuse of notation we define the representation map $R((v_1, v_2, v_3))$ as the piecewise linear function whose graph corresponds to a triangle with vertex on t_2 and base with extremal points the unique couple $a < b$ such that

$$\int_a^{v_1} [R((v_1, v_2, v_3))] (u)du = t_1, \qquad \int_{v_3}^b [R((v_1, v_2, v_3))] (u)du = 1 - t_3.$$

It is called triangular distribution associated to the triplet of quantile values (v_1, v_2, v_3).

A small abuse of notation consists in confusing the mode of the triangle with its median, losing the fundamental property of the representation maps, but in practice most of the experts in their mind do this confusion that moreover is not significant in distributions that are not too much skewed (the median and the mean are shifted with respect to the mode in case of major skewness). Another issue that come with this representation map is the automatical definition of the range of the distribution, and this could become a problem if the range is imposed a priori, like in the case of the elicitation of a probability value that must be in [0, 1]. In that case we put all the mass that falls outside the intrinsic range as a Dirac δ on the associated extremum (other solutions are reasonable, as a uniform spreading of that mass on the interval next to the extremum). To use this representation map in Monte Carlo simulations we implemented a very fast Newton-Raphson algorithm (reported in Chapter 6) to invert the explicit expression of the cumulative distribution of such a triangular pdf as a function of (a, b).

5.2.5. Pooling rules comparison

We give three different examples of the implications of adopting linear pooling or quantile pooling rules, obtained with $e = 1, 2$, $w = (0.5, 0.5)$ and both f_1 and f_2 are gaussian. The consequences of a different mean (Figure 5.2), a different standard deviation (sd) (Figure 5.3), and both at the same time (Figure 5.4), are explored. It is obtained that:

- the median of the quantile pooled pdf is the weighted combination of the two initial median values; instead the mean of the linear pooled pdf is the weighted combination of the two initial mean values.
- the shape of quantile pooled pdf (or cdf) is still gaussian, and has a sd that is the weighted combination of the two initial sd (indeed the sd coincides with a particular quantile).
- the shape of linear pooled pdf is bimodal on the two initial mean values, and the second moment is the weighted combination of the two initial second moments: hence the sd can be very large, including also an effect from the difference between the initial mean values. Even in case of equal initial mean values, the DM sd can be larger than the weighted combination of initial sd (due to the nonlinear operation of square root).

In summary, the quantile pooling results in a distribution that preserves the average width of the uncertainty bounds given by the single experts, and is centered on the average of their central estimates. This could correspond to focus first on the single median values for calculating the DM median, and then to transport the average uncertainty onto this.

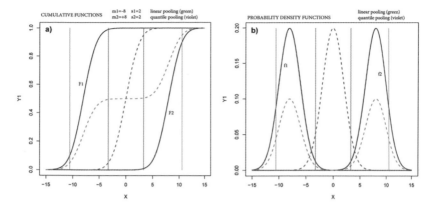

Figure 5.2. Linear (green line) vs quantile (violet line) pooling rules comparison, assuming two experts with equal weight and uncertainty distributions gaussian $\mathcal{N}(-8, 2)$ (red line) and $\mathcal{N}(8, 2)$ (blue line) respectively. In (a) are reported cumulative functions and in (b) probability density functions; in both cases also probability percentiles are plotted.

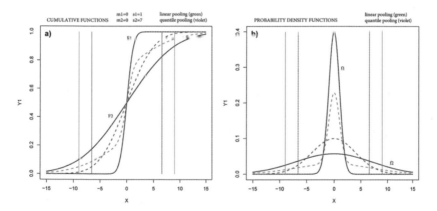

Figure 5.3. Linear (green line) vs quantile (violet line) pooling rules comparison, assuming two experts with equal weight and uncertainty distributions gaussian $\mathcal{N}(0, 1)$ (red line) and $\mathcal{N}(0, 7)$ (blue line) respectively. In (a) are reported cumulative functions and in (b) probability density functions; in both cases also probability percentiles are plotted.

In other words with a quantile pooling it is assumed to believe in the average of the experts' best guesses, affected by an averaged uncertainty. Conversely, the linear pooling combines the differences between the experts as additional uncertainty, preserving local modes in correspondence of their precise answers. The resulting distribution does not lose any information, reaching a comprehensive view of the uncertainties in play

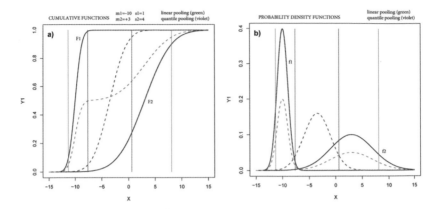

Figure 5.4. Linear (green line) vs quantile (violet line) pooling rules comparison, assuming two experts with equal weight and uncertainty distributions gaussian $\mathcal{N}(-10, 1)$ (red line) and $\mathcal{N}(3, 4)$ (blue line) respectively. In (a) are reported cumulative functions and in (b) probability density functions; in both cases also probability percentiles are plotted.

(even a lot wider than each initial single uncertainty), and it is possible to obtain cases in which the mean value (that is the weighted average of initial mean values) does not get any likelihood.

In practice the decision of assuming a linear pooling corresponds to choose randomly (in proportion the weights) an expert and believe to him: the variability of the DM being the combination of the variability between experts and of single experts. In conclusion the linear pooling gives a mixed DM that expresses the total uncertainty in play, without any statistic, while the quantile pooling presents a DM ruled by a statistic of the median and of one of the standard deviation.

5.3. The box model dynamic system and its solution

In fluid dynamics, a gravity current occurs whenever fluid of one density flows primarily horizontally in a gravitational field into fluid of a different density, hence gravity currents are sometimes also called 'density currents'. Gravity currents arise frequently in industrial, laboratory and natural situations. Examples in nature include avalanches, dust storms, seafloor turbidity currents, lahars, pyroclastic flows, and lava flows. A typical gravity current consists of a head and tail structure. The head, which is the leading edge of the gravity current, is a region in which relatively large volumes of ambient fluid are displaced. The tail is the bulk of fluid which follows the head. Immediately after the head, intense mixing occurs between the gravity current and the ambient fluid. Turbulent billows (Kelvin-Helmholtz instabilities) form in the wake of the head and engulf ambient fluid into the tail, a process called 'entrainment'.

The first analysis of the motion of a gravity current was carried out by von Kármán in 1940, in response to an enquiry by the American military before World War II concerning the wind speeds that would blow released nerve gas back onto friendly troops. Gravity currents can be either finite in volume, such as the release from a dam break, or continuously supplied from a source, such as in doorway. In the case of constant releases, the fluid in the head is constantly replaced and the gravity current is in some approximation stationary in the time window considered, but most gravity currents related to natural hazards anyways occur as a result of a finite-volume release of fluid and a transient dynamics. In the case where the widths of the initial release and of the environment are the same, one obtains what is usually referred to as a 'lock-exchange' flow. This refers to the flow spreading along walls on both sides and effectively keeping a constant width whilst it propagates. In this case the flow is effectively two-dimensional. Otherwise, if the flow spreads radially from the source forms an 'axisymmetric' flow. In the case of a point release, an extremely rare event in nature, the spread is perfectly axisymmetric, while in all other cases the current will form a sector, with the angle of spread depending on the release conditions.

When a gravity current encounters a solid boundary, it can either overcome the boundary by flowing around or over it, or be reflected by it. The actual outcome of the collision depends primarily on the height and width of the obstacle. If the obstacle is shallow (part) of the gravity current will overcome the obstacle by flowing over it. Similarly, if the width of the obstacle is small, the gravity current will flow around it, just like a river flows around a boulder. In 1940 von Kármán empirically established a law about the currents intruding along an horizontal base beneath a very deep otherwise quiescent fluid, obtained from the Bernoulli equation for inviscid flows; the law was made rigorous by Benjamin in 1968. The velocity of the front of the current u is related to the depth h of the current just behind the head by

$$u = Fr \left(g'h\right)^{1/2}$$

where $g' := \frac{\rho_c - \rho_a}{\rho_a} g$ is the reduced gravity acceleration defined in terms of the densities ρ_c and ρ_a of the current and ambient respectively, and Fr is the Froude number, $i.e.$ the square root of the ratio between inertial and gravity forces: the greater is the Froude number, the greater is the inertial resistance.

For a finite volume gravity current, perhaps the simplest modelling approach is via a box model, where a 'box' (rectangle for 2D problems,

cylinder for 3D) is used to represent the current. The box does not ro-
tate or shear, but changes in aspect ratio (*i.e.* stretches out) as the flow
progresses. Here, the dynamics of the problem are greatly simplified (*i.e.*
the forces controlling the flow are not direct considered, only their effects) and typically reduce to a condition dictating the motion of the front
via a Froude number and an equation stating the global conservation of
mass. The model is not a good approximation in the early slumping stage
of a gravity current, where h along the current is not at all constant, or in
the final viscous stage of a gravity current, where friction becomes important and changes Fr. The model is a good in the stage between these,
where the Froude number at the front is constant and the shape of the current has a nearly constant height. Additional equations can be specified
for processes that would alter the density of the intruding fluid such as
through sedimentation. Let set $u(t)$ the velocity of the front of the current, and $l(t)$ its position as a function of time. For simplicity this first
example does not model the deposition of mass. Entrainment of ambient
fluid is assumed negligible, and the flow instantaneously released and of
fixed volume.

Rectangular geometry Consider a two dimensional or 'lock exchange'
flow (with rectangular 'cartesian' geometry)

$$u = \frac{dl}{dt} = Fr(g'h)^{1/2},$$
$$lh = A,$$

where A the constant volume per unit width, or two dimensional area.
The equation

$$l^{1/2}dl = Fr(g'A)^{1/2}dt = dt/\tau,$$

where $\tau = (Fr^2 g'A)^{-1/2}$ represents the time scale of the dynamics, can
be integrated with initial condition $l(0) = 0$ to the function $\frac{2}{3}l^{3/2} = t/\tau$,
and inverted as

$$l(t) = \left(\frac{3}{2}t/\tau\right)^{2/3}.$$

Cylindrical geometry Assume a radially spreading flow: then instead of
the bidimensional condition $lh = A$ it satisfies

$$l^2 h = V$$

where V the volume of collapsing mixture divided by π (it is possible to
consider currents spreading on circular sectors). Again without deposi-

tion of mass, the equation becomes

$$dl = Fr \left(g_p \frac{V}{l^2} \right)^{1/2} dt = l^{-1} dt / \tau,$$

where the time scale is again of the form $\tau = (Fr^2 g'V)^{-1/2}$. Integrating the equation $l\, dl = dt/\tau$ as $\frac{1}{2}l^2 = t/\tau$ the solution is

$$l(t) = \sqrt{2t/\tau}.$$

Remark 5.26 (Shallow water equations). The Saint Venant equations, also called shallow water equations, are derived from depth-integrating the Navier-Stokes equations, in the case where the horizontal length scale is much greater than the vertical length scale, and they describe the flow below a pressure surface in a fluid. Using the same geometric relations it is possible to obtain similarity solutions to them that are identical to the box model solutions, but with a different multiplicative constant

$$\left(\frac{27 Fr^2}{12 - 2Fr^2} \right)^{1/3}.$$

5.3.1. The deposition of mass

Suspended particles are fundamentally different from homogeneous single phase flows, because the heavy particles continually settle down to the ground. The concentration of particles thus changes with time and position, and the density ρ_c can be written as

$$\rho_c = \rho_p \phi + \rho_a (1 - \phi),$$

where ρ_p is the density of the particles (the density of the interstitial fluid has been assumed equal to that of the ambient). In this case the reduced gravity g' is expressed as $g' = g_p \phi$, where

$$g_p = \frac{\rho_p - \rho_a}{\rho_a} g,$$

so the front condition becomes

$$\frac{dl}{dt} = Fr \left(g_p \phi h \right)^{1/2}.$$

The evolution of particle concentration in the flow is governed by

$$\frac{d\phi}{dt} = -w_s \frac{\phi}{h},$$

where w_s is the Stokes free-fall velocity given by

$$w_s = 2g_p a^2/(9\nu)$$

with ν the kinematic viscosity and a the length scale of the particle.

Remark 5.27 (Sediment erosion term). In principle it is also possible to add a term F to the equation

$$\frac{\phi}{dt} = -w_s\frac{\phi}{h} + F$$

representing the total vertical turbulent flux of sediment at the bed and the capacity of the flow to maintain sediment in suspension or rework newly deposed sediment. That sediment erosion term F should become negligible within the body of the flow if the ratio w_s/u is greater than $\sin\beta$ where β is the slope of the bed over which the current propagates.

Remark 5.28 (Interstitial fluid). It is possible to assume the fluid between the particles to have a different density ρ_i from the ambient fluid density ρ_a: in such case the reduced gravity takes a different form

$$g' = \frac{\rho_p\phi + \rho_i(1-\phi) - \rho_a}{\rho_a}g$$
$$= \left(\phi\frac{\rho_p - \rho_i}{\rho_a} + \frac{\rho_i - \rho_a}{\rho_a}\right)g,$$

and if are defined $\phi_{cr} := \frac{\rho_a - \rho_i}{\rho_p - \rho_i}$ and $g_{p,i} := \frac{\rho_p - \rho_i}{\rho_a}$ then

$$g' = g_{p,i}(\phi - \phi_{cr}).$$

The parameter ϕ_{cr} is the critical ratio of the density difference between the ambient and the interstitial fluids to that between the particles and the interstitial fluid. When $\phi_{cr} = 0$ it is $\rho_i = \rho_a$ and the standard case. If a gravity current with light interstitial fluid propagates below a denser ambient fluid, it is $\phi_{cr} > 0$ and when $\phi = \phi_{cr}$ the flow becomes neutrally buoyant and a further loss of driving buoyancy will result in the lift off of the current from the bed.

Rectangular geometry Let first assume again the 'lock exchange' fixed volume condition $lh = A$, obtaining formally:

$$\frac{d\phi}{dl} = \frac{d\phi}{dt}\frac{1}{\frac{dl}{dt}} = -w_s\frac{\phi}{h}\frac{1}{Fr\left(g_p\phi h\right)^{1/2}}$$
$$= -w_s\frac{\phi}{A}l\frac{l^{1/2}}{Fr\left(g_p\phi A\right)^{1/2}} = -\lambda l^{3/2}\phi^{1/2},$$

where

$$\lambda := \frac{w_s}{Fr\ g_p^{1/2} A^{3/2}}.$$

Consider the relation

$$\phi^{-1/2} d\phi = -\lambda l^{3/2} dl,$$

integrating it with starting conditions $\phi(0) = \phi_0, l(0) = 0$ gives

$$\phi^{1/2} = \phi_0^{1/2} - \frac{1}{5}\lambda l^{5/2}.$$

Impose $\phi = 0$ to find when the current ceases:

$$l_{max} = \left(\frac{5\phi_0^{1/2}}{\lambda}\right)^{2/5} = \left(\frac{5\phi_0^{1/2} Fr\ g_p^{1/2} A^{3/2}}{w_s}\right)^{2/5}.$$

Defining the non-dimensional variables $\Phi := \frac{\phi}{\phi_0}$ and $\xi := \frac{l}{l_{max}}$ it is possible to rewrite it as

$$\Phi = \left(1 - \xi^{5/2}\right)^2.$$

The front condition in non-dimensional variables, substituting the fixed volume condition in rectangular geometry is

$$d\xi = Fr\left(g_p \phi_0 A\right)^{1/2} l_{max}^{-3/2} \left(\frac{1 - \xi^{5/2}}{\xi^{1/2}}\right) dt = \left(\frac{1 - \xi^{5/2}}{\xi^{1/2}}\right) dt/\tau,$$

where

$$\tau = \left(Fr^2 g_p \phi_0 A\right)^{-1/2} l_{max}^{3/2}.$$

Setting

$$\mathcal{F}(\xi) := \int_0^\xi \frac{s^{1/2}}{1 - s^{5/2}} ds = t/\tau,$$

gives $\xi = \mathcal{F}^{-1}(t/\tau)$.

Cylindrical geometry Assuming again an axisymmetrical spread, the equation with deposition of mass becomes

$$\frac{dl}{dt} = Fr(g_p \phi V)^{1/2} l^{-1},$$

in which it is substituted like above $\frac{\phi}{dt} = -w_s \frac{\phi}{h}$. With the fixed volume condition $l^2 h = V$, it is formally:

$$\frac{d\phi}{dl} = \frac{d\phi}{dt}\frac{1}{\frac{dl}{dt}} = -w_s \frac{\phi}{h}\frac{1}{Fr\,(g_p\phi h)^{1/2}}$$

$$= -w_s \frac{\phi}{V} l^2 \frac{l}{Fr\,(g_p\phi V)^{1/2}} = -\lambda l^3 \phi^{1/2},$$

where λ is the same constant defined above, with V in place of A. Integrating the relation

$$\phi^{-1/2}d\phi = -\lambda l^3 dl$$

with starting conditions $\phi(0) = \phi_0, l(0) = 0$, it is obtained

$$\phi^{1/2} = \phi_0^{1/2} - \lambda l^4/8.$$

It is imposed $\phi = 0$ to find when the current ceases:

$$l_{max} = \left(\frac{8\phi_0^{1/2}}{\lambda}\right)^{1/4} = \left(\frac{8\phi_0^{1/2}Fr\,g_p^{1/2}V^{3/2}}{w_s}\right)^{1/4}.$$

If the non-dimensional variables Φ and ξ are considered (see above), then

$$\Phi = \left(1 - \xi^4\right)^2.$$

The front condition in non-dimensional variables, substituting the fixed volume condition, is

$$d\xi = Fr\,(g_p\phi_0 V)^{1/2} l_{max}^{-2}\left(\frac{1-\xi^4}{\xi}\right)dt = \left(\frac{1-\xi^4}{\xi}\right)dt/\tau,$$

where in this case

$$\tau = \left(Fr^2 g_p\phi_0 V\right)^{-1/2} l_{max}^2.$$

Let set

$$\mathcal{F}(\xi) := \int_0^\xi \frac{s}{1-s^4}ds = t/\tau,$$

hence

$$\mathcal{F}(\xi) = \frac{1}{4}\log\left(\frac{1+\xi^2}{1-\xi^2}\right),$$

so $e^{4t/\tau} = \frac{1+\xi^2}{1-\xi^2}$. If is considered the function

$$\xi(t) = (\tanh(2t/\tau))^{1/2} = \left(\frac{e^{2t/\tau} - e^{-2t/\tau}}{e^{2t/\tau} + e^{-2t/\tau}}\right)^{1/2},$$

then it is: $\xi = \mathcal{F}^{-1}(t/\tau)$ and

$$l(t) = [\tanh(2t/\tau)]^{1/2} \, l_{\max}$$

solves the equation.

Remark 5.29. From the expression of $l_{\max}$ in terms of the physical parameters is possible to write a formula for the area A or the volume V of the flow. It is given that

$$l_{\max} \propto A^{3/5}$$

and in particular

$$A = \left(\frac{l_{\max}^5 \, w_s^2}{\phi_0 g_p Fr^2} \right)^{1/3} / 5^{2/3}$$

assuming rectangular geometry, while

$$l_{\max} \propto V^{3/8}$$

and

$$V = \left(\frac{l_{\max}^8 \, w_s^2}{\phi_0 g_p Fr^2} \right)^{1/3} / 4$$

assuming cylindrical geometry. This estimate of the volume permits to state a simple estimate of the mass: $M = \rho_c V$, where $\rho_c = \phi_0 \rho_p + (1 - \phi_0)\rho_a$.

5.3.2. The energy function

The 'box model' approximation permits a first approximation of the kinetic energy of the front of the flow as a function of the distance l; following the idea behind the energy line approach, it is easy to compare it with the potential energy associated to overcoming topographical barriers:

$$\frac{1}{2}u^2 = gH$$

where H is the height of the barrier, the mass of the flow front is simplified, and simple gravity g is considered, also neglecting hydraulic effects associated with flow-obstacle interactions. It is given that

$$H = \frac{1}{2g}u^2,$$

where u is the front velocity.

Rectangular geometry Besides for box model dynamical system with rectangular geometry it is not possible to exactly integrate the differential equation, anyways an expression for the decay of the kinetic energy is obtained deriving its implicit integral form:

$$u(t) = \frac{l_{\max}}{\tau \mathcal{F}' \left(\mathcal{F}^{-1}(t/\tau) \right)},$$

but $\mathcal{F}^{-1}(t/\tau) = \frac{l}{l_{\max}}$ and $\mathcal{F}'(s) = \frac{s^{1/2}}{1 - s^{5/2}}$ hence

$$u(t(l)) = \left(\frac{1 - x^{5/2}}{x^{1/2}} \right) l_{\max}/\tau,$$

where $x = \frac{l}{l_{\max}}$. In this case the decay function is

$$H = \frac{1}{2g} \left[l_{\max}/\tau \frac{1 - x^{5/2}}{x^{1/2}} \right]^2, \tag{5.1}$$

and it is also possible to simplify the equation, leaving an implicit dependence on A only through $l_{\max}$:

$$l_{\max}/\tau = (1/5)^{1/3} \left(w_s Fr^2 \phi_0 g_p \right)^{1/3} l_{\max}^{1/3} = (8/5)^{1/3} C \, l_{\max}^{1/3},$$

where

$$C := \left(w_s Fr^2 \phi_0 g_p \right)^{1/3} /2.$$

Cylindrical geometry In this case it is possible to derive explicitly $l(t)$ obtaining that

$$u(t) = \frac{l_{\max} \, (\tanh(2t/\tau))^{-1/2}}{\tau \cosh^2(2t/\tau)},$$

but $t = \frac{\text{artanh}}{2\tau} \left[\left(\frac{l}{l_{\max}} \right)^2 \right]$, hence

$$u(t(l)) = \frac{l_{\max}^2}{\tau l \cosh^2 \left(\text{artanh} \left[\left(\frac{l}{l_{\max}} \right)^2 \right] \right)}$$

$$= \frac{l_{\max}}{\tau x \cosh^2 \text{artanh}(x^2)},$$

where $x = \frac{l}{l_{\max}}$. The function

$$H = \frac{1}{2g} \left[\frac{l_{\max}}{\tau x \cosh^2 \text{artanh}(x^2)} \right]^2 \tag{5.2}$$

replaces the straight line of the classical energy cone model (note that this one diverges in zero). Also in this case it is possible to simplify the expression:

$$l_{max}/\tau = (1/8)^{1/4} \left(w_s^2 g_p^3 \phi_0^3 F r^6 V \right)^{1/8}$$
$$= (1/8)^{1/3} \left(w_s F r^2 \phi_0 g_p \right)^{1/3} l_{max}^{1/3} = C \, l_{max}^{1/3},$$

where C has the same meaning as in the rectangular case.

Remark 5.30. Both the two functions $\frac{1}{x \cosh^2 \operatorname{artanh}(x^2)}$ and $\frac{1-x^{5/2}}{x^{1/2}}$ are decreasing in x so increasing as functions of l_{max} once is fixed a distance l from the vent; hence in both the cases $H(l, l_{max})$ is an increasing function of l_{max} for each $l > 0$ and so also the invaded area obtained by energy comparison is an increasing function of it. This is a key property for inverting such function (in general not even continuous) and obtaining an estimate of l_{max}, hence of the volume, from the areal size invaded.

5.4. The doubly stochastic Hawkes processes

In many contexts, the observations of a system are represented as events associated with elements of a given space, that arrive randomly through time but are not stochastically independent. Chapter 4 was focused on the simulation of the eruptive activity in a volcanic field with multivariate counting processes capable of re-producing clusters in time-space (see also [15]). The Hawkes process structure was adopted for implementing a self-excitement behavior: such class of processes finds wide applications in many applied fields, including seismology for modelling earthquakes replicas, neuroscience for modelling brain activity, genome analysis, financial contagion, market micro-structure, social networks interactions, epidemiology (see [49] for some references).

In addition, the large uncertainty that affects the available data forced us to quantify it and explore how it may affect the model. A well known class of processes capable to take into account an uncertainty source on the parameters is called the Cox processes, or doubly stochastic Poisson processes (see also [88, 89, 90] for examples in volcanology). Hence we developed a time space mathematical model that takes into account both self-exciting and uncertainty sources: we called it a Cox-Hawkes process. In the following we explain what a time non-homogeneous Poisson process with a given intensity function is, focusing on the Hawkes processes and giving the maximum likelihood (ML) expressions for the main parameters.

5.4.1. Cox processes

Definition 5.31 (Counting process). Let $(\Omega, \mathcal{F}, (F_t)_{t\geq 0}, P)$ be a filtered probability space. We say that $Z = (Z_t)_{t\geq 0}$ is a counting process if it is non-decreasing, càdlàg, integer valued (and finite all times), with all its jumps of height 1.

This wide class includes the Poisson processes (homogeneous and non-homogeneous), the renewal processes and even non-stationary processes. The counting processes are representable as special point processes (see *e.g.* [44]), assuming them as random measures supported on the points of the real positive line.

Definition 5.32 (Point process). A boundedly finite measure is a measure μ on the Borel family of a complete separable metric space $\mathcal{X}$ such that $\mu(A) < +\infty$ for all $A \in \mathcal{B}(\mathcal{X})$ that is bounded. A point process N is a boundedly finite and integer valued random measure. A simple point process is a point process N^* such that a.s. for each $x \in \mathcal{X}$ verifies $N\{x\} = 0$ or 1, and we have that $N^* = \sum_i \delta_{x_i}$, where δ_{x_i} are the Dirac-delta for some family of points $x_i \in \mathcal{X}$

In particular each counting process Z can be associated to a simple point process N^* on $\mathbb{R}_+$: for each interval $[0, t]$ we define $N^*([0, t]) = Z_t$ and extend the measure to the Borel σ-algebra. Moreover, the same relation define uniquely a counting process Z for each given simple point process N^* on $\mathbb{R}_+$ (see [46]). In the sequel we will define Cox and Hawkes as counting processes, but without loss of generality we can assume them as point processes. The average density of points of a counting process is represented by its intensity function.

Proposition 5.33 (Intensity function). *Let Z be a $(\mathcal{F}_t)$-adapted counting process. There is a unique non-decreasing predictable process $(\Lambda_t)_{t\geq 0}$, called compensator of Z, such that $(Z_t - \Lambda_t)_{t\geq 0}$ is a $(\mathcal{F}_t)_{t\geq 0}$-local martingale. The (distributional) time derivative $(\lambda_t)_{t\geq 0}$ of the compensator is called intensity function (or measure) of Z (see [46, 87]).*

The homogeneous Poisson processes assume independent identically distributed exponential time intervals between the events: they are ruled by a constant intensity parameter λ directly representing the mean number of points extracted in each unit of time. Taking into account deterministic locally integrable intensity functions $\lambda(t)$ and still assuming exponentially distributed renewal times, leads to the class of non-homogeneous Poisson processes (NHPP). Sometimes this class is assumed to include also the spatially non-homogeneous Poisson processes, which are point processes on a space-time domain.

Definition 5.34 (Non-homogeneous Poisson process). Given a non-negative function λ defined on $\mathbb{R}_+$ and such that $\lambda \in \mathcal{L}^1([0, T])$ for all $T > 0$, we say that a counting process Z is a NHPP of intensity $\lambda(\cdot)$ if it has independent increments and for all $t > s$ it satisfies

$$P(Z_t - Z_s = n) = \exp(-\Lambda(s, t))\frac{\Lambda(s, t)^n}{n!},$$

where $\Lambda(s, t) = \int_s^t \lambda(u)du$. The function $\Lambda_t := \Lambda(0, t)$ is the compensator of the Poisson process (it is easy to prove that the process $Z - \Lambda$ is a martingale).

An immediate consequence of the definition is the trivial existence of such processes: if we take Z a homogeneous Poisson process with intensity equal to 1 and Λ a continuous increasing function defined on $\mathbb{R}_+$ such that $\Lambda(0) = 0$, then the process $Y_t = Z_{\Lambda(t)}$ is a NHPP with compensator Λ. This principle implies the following result.

Lemma 5.35 (Sample space). *Let Z be a homogeneous Poisson process with intensity equal to 1, that is defined on the probability space $(\Omega, \mathcal{F}, P)$. Then every NHPP may be defined on $(\Omega, \mathcal{F}, P)$.*

We also state a fundamental principle that will be useful to simulate NHPPs from the outcomes of an assigned homogeneous Poisson process, this is called the thinning property (see [46]).

Proposition 5.36 (Thinning property). *Let N be a NHPP with intensity $\lambda(\cdot)$ and let $I = (\tau_i)_{i=1,\ldots,n}$ the instants of its first n points in $[0, T]$. Let λ' is a positive function such that $\lambda'(t) < \lambda(t)$ in $[0, T]$. Then we define a subfamily $I' \subseteq I$ eliminating the instant τ_i with probability $1 - \lambda'(\tau_i)/\lambda(\tau_i)$. Then I' contains the instants on the interval $[0, T]$ of the first points of a NHPP with intensity $\lambda'(\cdot)$.*

A more general class of NHPP is obtained assuming the intensity functions as (nonnegative) random variables themselves: both the Cox processes and the Hawkes processes belong to this class (see [44, 46]).

Definition 5.37 (Cox process). Let Z be a counting process with a random intensity function $\lambda(t, \omega)$. Assume that Z conditional on each trajectory of λ, *i.e.* $Z|\lambda$, is a NHPP. Then Z is called a Cox process, or doubly stochastic Poisson process.

In particular the intensity can always be sampled before sampling the counting process. The existence of such a process once given an arbitrary intensity process λ is straightforward on a probability space $\Omega := \Omega_1 \times \Omega_2$, where on the first space we extract the random intensity and then on the second space we extract the trajectory as a NHPP of that intensity; we remark that using Lemma 5.35 the space Ω_2 can be fixed a priori.

5.4.2. Self-interacting processes

If we do not assume the independence of a random intensity from the previously observed events of the counting process, we can implement a self-interaction behavior. Assuming to take an intensity function $\lambda(t, \omega)$ that is a predictable process with respect to the filtration, we include also the important class of Hawkes processes described in the following.

Definition 5.38 (Conditional intensity). Let Z be a counting process with a random intensity function $\lambda(t, \omega)$ which is a predictable process. A predictable process is always $\mathcal{F}_-$ measurable, so using the Doob lemma, we have that

$$\lambda(t, \omega) = \lambda^*(t, t_1(\omega), \ldots, t_{n(\omega)}(\omega)),$$

where $(t_i(\omega))_{0 < i < n(\omega)}$ is the family of the instants of the points of the process before t, and $\lambda^*(\cdot)$ is a function defined on the finite sequences of positive numbers. This function $\lambda^*(\cdot)$ is called the conditional intensity of Z.

Remark 5.39. The concept that is behind the general definition of the conditional intensity of a general point process is the Radon-Nikodym derivative of the so called Campbell measure associated. Let Z be a non-exploding counting process: then we consider the associated simple point process $N(\omega)$ defined above. If we take Ω as the metric space of all the boundedly finite measures on $\mathbb{R}_+$ with its Borel σ-algebra, and we define $U := \mathbb{R}_+ \times \Omega$, we can introduce the Campbell measure as the measure $N \otimes P$ on U. Then it is possible to prove that the conditional intensity λ of Z is the Radon-Nikodym derivative of this measure with respect to $\mathcal{L} \otimes P$ (where $\mathcal{L}$ is the Lebesgue measure); see more details in [46].

This leads to the possibility to define a priori the function λ^* for representing the behaviour of the process.

Definition 5.40 (Hazard functions). A counting process Z is called regular if for each sequence of instants $I = (t_1 < \cdots < t_{n-1})$ there exists a conditional density $p_n(t|t_1, \ldots, t_{n-1})$ associated to the probability distribution of the instant of the nth point after $n - 1$ points on the given times. We define $S_n(t|t_1, \ldots, t_{n-1})$ as the survival function for this probability measure. For all $n \in \mathbb{N}$ the nth hazard function of Z is

$$h_n(t|t_1, \ldots, t_{n-1}) := \frac{p_n(t|t_1, \ldots, t_{n-1})}{S_n(t|t_1, \ldots, t_{n-1})}.$$

In the simple case of a homogeneous Poisson process of intensity λ, we have

$$p_n(t|t_1, \ldots, t_{n-1}) = \lambda \exp\left(-\lambda(t - t_{n-1})\right),$$

$$S_n(t|t_1, \ldots, t_{n-1}) = \exp\left(-\lambda(t - t_{n-1})\right), \quad h_n(t|t_1, \ldots, t_{n-1}) = \lambda$$

for each family I of instants before t. Moreover in general from the definition $p_n = -\frac{d}{dt} S_n$ we obtain that

$$\int_{t_{n-1}}^{t} h_n(s|t_1, \ldots, t_{n-1}) ds = -\log\left(S_n(t|t_1, \ldots, t_{n-1})\right),$$

$$S_n(t|t_1, \ldots, t_{n-1}) = \exp\left(-\int_{t_{n-1}}^{t} h_n(s|t_1, \ldots, t_{n-1}) ds\right),$$

hence

$$p_n(t|t_1, \ldots, t_{n-1}) = h_n(s|t_1, \ldots, t_{n-1}) \exp\left(-\int_{t_{n-1}}^{t} h_n(s|t_1, \ldots, t_{n-1}) ds\right).$$

Using the hazard functions we can give a more general definition for the conditional intensity λ^* given a sequence of instants $I = (t_1 < \cdots < t_M)$:

$$\lambda^*(t|t_1, \ldots, t_M) = \begin{cases} h_1(t), & 0 < t \leq t_1; \\ h_n(t|t_1, \ldots, t_{n-1}), & t_{n-1} < t \leq t_n. \end{cases}$$

We remark that the choice of the hazard functions or of the conditional densities corresponds to assume a density for the Janossy measures, a family of symmetrical distributions on $\mathbb{R}_+^n$, characterizing uniquely the law of the counting process (see [46]). The following is a simple existence result for a counting process of given hazard functions $(h_n)_{n \geq 0}$. This is a local result: the counting process is not bounded in general and could still explode after an arbitrarily small time $\tau(\omega)$.

Theorem 5.41 (Local existence). *Let $(h_n)_{n \geq 0}$ be a family of functions such that $h_n : S_n \to \mathbb{R}$, where S_n is the set of strictly increasing n-uples of positive real numbers. We will follow the notation $h_n(t|t_1, \ldots, t_{n-1})$ where $(t_1 < \cdots < t_{n-1} < t) \in S_n$. Then there exists a possibly exploding counting process N such that its intensity is well defined and coincides with*

$$\lambda(t, \omega) = h_n(t|t_1(\omega), \ldots, t_{n-1}(\omega))$$

for each n,t such that $t_{n-1}(\omega) < t < t_n$, where we assumed $t_0 = 0$ for simplicity. The trajectory $N(\omega)$ will be finite at least for all the times in an interval $[0, \tau(\omega))$, where τ is a strictly positive random variable.

Proof. Let N^0 be a NHPP of intensity $\lambda_1(t) = h_1(t)$, defined on a probability space Ω_1. Let $\tau_1(\omega_1) = t_1^{N^0}(\omega_1)$ be the instant of its first point. For each $\omega_1 \in \Omega_1$ we define $N^1(\omega_1)$ on the space Ω_2, as a NHPP of intensity

$$\lambda_2^{\omega_1}(t) = h_2(t|\tau_1(\omega_1))$$

for all $t > \tau_1(\omega_1)$, and null otherwise. Let $\tau_2(\omega_1, \omega_2) = t_1^{N^1(\omega_1)}(\omega_2)$ be the instant of its first point. In general for each $n \in \mathbb{N}$ we assume a recursive definition for $N^n(\omega_1, \ldots, \omega_n)$ as a NHPP of intensity

$$\lambda_{n+1}^{\omega_1, \ldots, \omega_n}(t) = h_{n+1}(t|\tau_1(\omega_1), \ldots, \tau_n(\omega_1, \ldots, \omega_n))$$

for all $t > \tau_n(\omega_1, \ldots, \omega_n)$, and null otherwise. Using the notation $N^i|_{\tau_i}$ to call the process N^i arrested at a stopping time τ_i, we can define the counting process $N := \sum_{i=0}^{\infty} N^i|_{\tau_{i+1}}$, on the probability space $\Omega := \prod_{i=0}^{\infty} \Omega_i$. Its intensity verifies the hypotheses. The process in general does not explode before a time $\tau(\omega) = \sup_i \tau_i(\omega_i)$. Moreover, thanks to Lemma 5.35, each space Ω_i has not to depend on the previous extractions and can be fixed a priori. □

A multivariate counting process is a family $(Z_t^1, \ldots, Z_t^n)_{t \geq 0}$ of counting processes, and may be considered as a point process on $\mathbb{R}_+^n$. From a practical point of view each component Z_t^i records the number of events of the i-th component of a system during $[0, t]$ or equivalently the time stamps of the observed events. Under relatively weak assumptions (see [46]) a multivariate counting process is characterized by its intensity process $(\lambda_t^1, \ldots, \lambda_t^n)_{t \geq 0}$: this function is as usual defined as the (distributional) derivative of the compensator of Z and heuristically satisfies the differential expression

$$\lambda_t^i dt = P\left(Z^i \text{ has a jump in } [t, t + dt] | \mathcal{F}_t\right), \quad i = 1, \ldots, n$$

where $\mathcal{F}_t$ denotes the sigma-algebra generated by $(Z^i)_{1 \leq i \leq n}$ up to time t.

5.4.3. Hawkes processes

Assuming the most general multivariate framework, following [49] we consider a countable directed graph $G = (\mathcal{S}, \mathcal{E})$ with vertices $i \in \mathcal{S}$ and (directed) edges $e \in \mathcal{E}$. We write $e = (j, i) \in \mathcal{E}$ for the oriented edge. We also need to specify the following parameters: a kernel of causal functions $\varphi = (\varphi_{ij}, (i, j) \in \mathcal{E})$ with $\varphi_{ij} : [0, +\infty) \mapsto \mathbb{R}$ and a (possibly) nonlinear intensity component $f = (f_i, i \in \mathcal{S})$ with $f_i : \mathbb{R} \mapsto [0, +\infty)$. This is equivalent to impose the hazard functions of the process.

Definition 5.42. A multivariate Hawkes process with parameters (G, φ, f) is a family of $(\mathcal{F}_t)_{t \geq 0}$-adapted counting processes $(Z_t^i)_{i \in \mathcal{S}, t \geq 0}$ such that

(i) almost surely, for all $i \neq j$, $(Z_t^i)_{t \geq 0}$ and $(Z_t^j)_{t \geq 0}$ never jump simultaneously,

(ii) for every $i \in S$, the compensator $(\Lambda_t^i)_{t \geq 0}$ of $(Z_t^i)_{t \geq 0}$ has the form $\Lambda_t^i = \int_0^t \lambda_s^i ds$, where the intensity process $(\lambda_t^i)_{t \geq 0}$ is given by

$$\lambda_t^i = f_i \left(\sum_{j \to i} \int_0^t \varphi_{ij}(t - s) dZ_s^j \right),$$

with the notation $\sum_{j \to i}$ for summation over $\{ j : (j, i) \in \mathcal{E} \}$.

We say that a Hawkes process is linear when $f_i(x) = \mu_i + x$ for every $x \in \mathbb{R}$, $i \in S$, with $\mu_i \geq 0$ and when $\varphi_{ji} \geq 0$. This is the case of Chapter 4, and μ_i can be interpreted as a baseline Poisson intensity. In the degenerate case $\varphi_{ij} = 0$, we actually retrieve homogeneous Poisson processes.

A multivariate Hawkes process $(Z_t^i)_{i \in S, t \geq 0}$ with parameters (G, φ, f) behaves as follows. For each $i \in S$, the rate of jump of Z^i is, at time t,

$$\lambda_i(t) = f_i \left(\sum_{j \to i} \sum_{k \geq 1} \varphi_{ji}(t - T_k^j) 1_{\{T_k^j \leq t\}} \right),$$

where $(T_k^j)_{k \geq 1}$ are the jump times of Z^j. In other words, each time one of the Z^js has a jump, it excites its neighbors in that it increases their rate of jump (in the usual situation when f is increasing and φ positive). If φ is positive and decreases to 0 the influence of a jump decreases and tends to 0 as time evolves.

We will give an existence and uniqueness result for Hawkes processes reported from [49]. It is based on the classical idea (see also [22]) of writing the process as the solution of a system of Poisson driven SDEs and of finding a set of assumptions on the parameters under which we can prove the pathwise existence and uniqueness for this system of SDEs. Consider, on a filtered probability space $(\Omega, \mathcal{F}, (\mathcal{F}_t)_{t \geq 0}, P)$, a family $(\pi^i(dsdz), i \in S)$ of independent identically distributed $(\mathcal{F}_t)_{t \geq 0}$-Poisson measures with intensity measure $dsdz$ on $[0, +\infty) \times [0, +\infty)$.

Definition 5.43. A family $(Z_t^i)_{i \in S, t \geq 0}$ of càdlàg $(\mathcal{F}_t)_{t \geq 0}$-adapted processes is called a multivariate Hawkes process with parameters (G, φ, f) is a.s., for all $i \in S$, all $t \geq 0$

$$Z_t^i = \int_0^t \int_0^\infty 1_{\left\{ z \leq f_i \left(\sum_{j \to i} \int_0^{s^-} \varphi_{ji}(s - u) dZ_u^j \right) \right\}} \pi^i(dsdz).$$

This integral formulation is consistent with Definition 5.42.

Proposition 5.44. *A Hawkes process in the sense of Definition* 5.43 *is also a Hawkes process in the sense of Definition* 5.42. *Viceversa, for each Hawkes process* $Z = (Z^i_t)_{i\in S, t\geq 0}$ *in the sense of Definition* 5.42 *we can build (on a possibly enlarged space) a family of independent identically distributed Poisson measures* $(\pi^i(dsdz), i \in S)$ *such that Z is a Hawkes process in the sense of Definition* 5.43.

Proof. The first point is straightforward: for a Hawkes process $(Z^i_t)_{i\in S, t\geq 0}$ in the sense of Definition 5.43, for every $i \in S$, the compensator of Z^i is

$$\int_0^t \int_0^\infty 1_{\left\{z \leq f_i\left(\sum_{j\to i}\int_0^{s-}\varphi_{ji}(s-u)dZ^j_u\right)\right\}} dsdz$$

with equals to

$$\int_0^t f_i\left(\sum_{j\to i}\int_0^{s-}\varphi_{ji}(s-u)dZ^j_u\right)ds.$$

The independence of the Poisson random measures guarantees that for all $i \neq j$ the processes $(Z^i_t)_{t\geq 0}$ and $(Z^j_t)_{t\geq 0}$ a.s. never jump simultaneously. The vice versa is more delicate, and the proof can be found in [46]: it is based on the construction algorithm of Lewis-Shedler, extended by Ogata. □

Still following [49], the following set of assumptions will guarantee the well-posedness of Definition 5.43.

Definition 5.45. A set of parameters (G, φ, f) is regular if there are some non-negative constants $(c_i)_{i\in S}$, some positive weights $(p_i)_{i\in S}$ and a locally integrable function $\phi : [0, +\infty) \mapsto [0, +\infty)$ such that

(i) for every $i \in S$, every $x, y \in \mathbb{R}$, $|f_i(x) - f_i(y)| \leq c_i|x - y|$,

(ii) $\sum_{i\in S} f_i(0)p_i < +\infty$,

(iii) for every $s \in [0, +\infty)$, every $j \in S$, $\sum_{i,(i,j)\in\mathcal{E}} c_i p_i|\varphi_{ji}(s)| \leq p_j\phi(s)$.

We remark that if S is finite, the previous properties are satisfied (with $p_i = 1$ for all $i \in S$) as soon as f_i is Lipshitz continuous for all $i \in S$ and φ_{ji} is locally integrable for all $(j, i) \in \mathcal{E}$. We give here some technical results about convolution equations.

Lemma 5.46. *Let* $\phi : [0, +\infty) \mapsto \mathbb{R}$ *be locally integrable and let* $\alpha : [0, +\infty) \mapsto \mathbb{R}$ *have finite variations on compact intervals and satisfy*

$\alpha(0) = 0$. *Then for all* $t \geq 0$,

$$\int_0^t \int_0^{s-} \phi(s-u)d\alpha(u)ds = \int_0^t \int_0^s \phi(s-u)d\alpha(u)ds$$

$$= \int_0^t \phi(t-s)\alpha(s)ds.$$

Proof. First we clearly have that

$$\int_0^{s-} \phi(s-u)d\alpha(u) = \int_0^s \phi(s-u)d\alpha(u)$$

for almost every $s \geq 0$, whence we obtain the first equality. Using twice the Fubini theorem,

$$\int_0^t \left(\int_0^s \phi(s-u)d\alpha(u) \right) ds = \int_0^t \left(\int_u^t \phi(s-u)ds \right) d\alpha(u)$$

$$= \int_0^t \left(\int_0^{t-u} \phi(v)dv \right) d\alpha(u)$$

$$= \int_0^t \left(\int_0^{t-v} d\alpha(u) \right) \phi(v)dv$$

$$= \int_0^t \alpha(t-v)\phi(v)dv,$$

from which the conclusion follows, using the substitution $s = t - v$. □

Lemma 5.47. *Let* $\phi : [0, +\infty) \mapsto [0, +\infty)$ *be locally integrable and* $g : [0, +\infty) \mapsto [0, +\infty)$ *be locally bounded.*

(i) *Consider a locally bounded non-negative function u such that*

$$u_t \leq g_t + \int_0^t \phi(t-s)u_s ds$$

for all $t \geq 0$. *Then* $\sup_{[0,T]} u_t \leq C_T \sup_{[0,T]} g_t$, *for some constant* C_T *depending only on* $T > 0$ *and* ϕ.

(ii) *Consider a sequence of locally bounded non-negative functions* u^n *such that*

$$u_t^{n+1} \leq \int_0^t \phi(t-s)u_t^n ds,$$

for all $t \geq 0$, *all* $n \geq 0$. *Then* $\sup_{[0,T]} \sum_{n \geq 0} u_t^n \geq C_T$, *for some constant* C_T *depending only on* $T > 0$, u^0 *and* ϕ.

(iii) *Consider a sequence of locally bounded non-negative functions* u^n *such that*

$$u_t^{n+1} \le g_t + \int_0^t \phi(t-s)u_s^n ds,$$

for all $t \ge 0$, *all* $n \ge 0$. *Then* $\sup_{[0,T]} \sup_{n \ge 0} u_t^n \le C_T$, *for all* $T \ge 0$ *and for some constant* C_T *depending only on* $T > 0$, u^0, g *and* ϕ.

Proof. We start with point (i). Fix $T > 0$ and consider $A > 0$ such that $\int_0^T \phi(s)1_{\{\phi(s)\ge A\}} ds \le 1/2$. Then for all $t \in [0,T]$,

$$u_t \le g_t + \int_0^t 1_{\{\phi(t-s)\le A\}}\phi(t-s)u_s ds + \int_0^t 1_{\{\phi(t-s)> A\}}\phi(t-s)u_s ds$$

$$\le g_t + A\int_0^t u_s ds + \sup_{[0,t]} u_s/2,$$

from which we deduce that

$$\sup_{[0,t]} u_s \le 2\sup_{[0,t]} g_s + 2A\int_0^t u_s ds.$$

We then can apply the standard Grönwall Lemma to get

$$\sup_{[0,T]} u_s \le 2\left(\sup_{[0,T]} g_s\right)e^{2AT}.$$

To check point (iii), put $v_t^n = \sup_{k=0,\dots,n} u_t^k$. One easily checks that

$$v_t^n \le u_t^0 + g_t + \int_0^t \phi(t-s)v_s^n ds,$$

for all $n \ge 0$. By point (i),

$$\sup_{[0,T]} v_t^n \le C_T \sup_{[0,T]}(g_t + u_t^0).$$

Letting n increase to infinity concludes the proof. Point (ii) follows from point (iii), since $v_t^n = \sum_{k=0}^n u_t^k$ satisfies

$$v_t^{n+1} \le u_t^0 + \int_0^t \phi(t-s)v_s^n ds. \qquad \square$$

The proof of the following theorem is based on a Cauchy approximation of the solution, thanks to Lemma 5.46 and 5.47.

Theorem 5.48. *If the set of parameters (G, φ, h) is regular, then exists a pathwise unique multivariate Hawkes process $(Z_t^i)_{i \in \mathcal{S}, t \geq 0}$ such that $\sum_{i \in \mathcal{S}} p_i E[Z_t^i] < +\infty$ for all $t \geq 0$.*

Proof. We first prove uniqueness. Let thus $(Z_t^i)_{i \in \mathcal{S}, t \geq 0}$ and $(\tilde{Z}_t^i)_{i \in \mathcal{S}, t \geq 0}$ be two solutions satisfying the conditions of Definition 5.43. Set

$$\Delta_t^i := \int_0^t |d(Z_s^i - \tilde{Z}_s^i)|, \quad \forall i \in \mathcal{S}, t \geq 0,$$

that is the total variation norm of the signed measure $d(Z_s^i - \tilde{Z}_s^i)$ on $[0, t]$. We also put $\delta_t^i := E[\Delta_t^i]$ and first prove that

$$\delta_t^i \leq c_i \int_0^t \sum_{j \to i} |\varphi_{ji}(t - s)| \delta_s^j ds.$$

Indeed we have

$$\Delta_t^i = \int_0^t \int_0^\infty \left| 1_{\left\{z \leq f_i\left(\sum_{j \to i} \int_0^{s-} \varphi_{ji}(s-u) dZ_u^j\right)\right\}} - 1_{\left\{z \leq f_i\left(\sum_{j \to i} \int_0^{s-} \varphi_{ji}(s-u) d\tilde{Z}_u^j\right)\right\}} \right|$$
$$\times \pi^i(ds dz).$$

Taking expectations, we deduce that

$$\delta_t^i = \int_0^t E\left[\left| f_i\left(\sum_{j \to i} \int_0^{s-} \varphi_{ji}(s - u) dZ_u^j\right) \right. \right.$$
$$\left. \left. - f_i\left(\sum_{j \to i} \int_0^{s-} \varphi_{ji}(s - u) d\tilde{Z}_u^j\right) \right| \right] ds$$
$$\leq c_i \sum_{j \to i} E\left[\int_0^t \int_0^{s-} |\varphi_{ji}(s - u)| d\Delta_u^j ds\right]$$

by Definition 5.45-(i). Using Lemma 5.46 we see that

$$\int_0^t ds \int_0^{s-} |\varphi_{ji}(s - u)| d\Delta_u^j = \int_0^t |\varphi_{ji}(t - u)| d\Delta_u^j du$$

which, plugged into the previous expression yields the wanted inequality. Set $\delta_t := \sum_{i \in \mathcal{S}} p_i \delta_t^i$, where the weights p_i were introduced in Definition 5.45-(ii). By assumption, δ_t is well defined and finite. We infer by the expression just proved that

$$\delta_t \leq \int_0^t \sum_{i \in \mathcal{S}} p_i c_i \sum_{j \to i} |\varphi_{ji}(t - s)| \delta_s^j ds.$$

By Definition 5.45-(iii),

$$\delta_t \leq \int_0^t \sum_{j \in \mathcal{S}} \sum_{i,(j,i) \in \mathcal{E}} c_i \, p_i |\varphi_{ji}(t-s)| ds$$

$$\leq \int_0^t \sum_{j \in \mathcal{S}} p_j \delta_s^j \phi(t-s) ds$$

$$= \int_0^t \phi(t-s) \delta_s de.$$

Lemma 5.47-(i) thus implies that $\delta_t \equiv 0$, from which uniqueness follows. We now prove existence. Let $Z_t^{i,0} := 0$ and , for $n \geq 0$,

$$Z_t^{i,n+1} := \int_0^t \int_0^\infty 1_{\{z \leq f_i(\sum_{j \to i} \int_0^{s-} \varphi_{ji}(s-u) dZ_u^{j,n})\}} \pi^i(dsdz).$$

We define

$$\delta_t^{i,n} := E\left[\int_0^t |dZ_s^{i,n+1} - dZ_s^{i,n}|\right]$$

and $\delta_t^n := \sum_{i \in \mathcal{S}} p_i \delta_t^{i,n}$. As in the proof of uniqueness, we obtain, for $n \geq 0$,

$$\delta_t^{n+1} \leq \int_0^t \phi(t-s) \delta_s^n ds.$$

Next, we put $m_t^{i,n} := E[Z_t^{i,n}]$. By Definition 5.45-(i), $f_i(x) \leq f_i(0) + c_i |x|$, whence

$$m_t^{i,n+1} \leq E\left[\int_0^t \left(f_i(0) + c_i \sum_{j \to i} \int_0^{s-} |\varphi_{ji}(s-u)| dZ_u^{j,n}\right) ds\right]_2$$

$$\leq \int_0^t \left(f_i(0) + c_i \sum_{j \to i} |\varphi_{ji}(t-s)| m_s^{j,n}\right) ds,$$

where we used that, by Lemma 5.46,

$$\int_0^t \int_0^{s-} |\varphi_{ji}(s-u)| dZ_u^{j,n} ds = \int_0^t |\varphi_{ji}(t-u)| Z_u^{j,n} du$$

Setting $u_t^n := \sum_{i \in \mathcal{S}} p_i m_t^{i,n}$ and using Definition 5.45-(ii,iii)

$$u_t^{n+1} \leq t \sum_{i \in \mathcal{S}} f_i(0) p_i + \int_0^t \sum_{i \in \mathcal{S}} p_i c_i \sum_{j \to i} |\varphi_{ji}(s-u)| m_s^{j,n} ds$$

$$\leq Ct + \int_0^t \phi(t-s) u_s^n ds.$$

Since $u_t^0 = 0$ and ϕ is locally integrable, by induction we have that u^n is locally bounded for all $n \geq 0$. Consequently, δ^n is also locally bounded for all $n \geq 0$. Lemma 5.47-(ii) implies that for all $T \geq 0$, $\sum_{n \geq 1} \delta_T^n < +\infty$. This classically implies that the sequence is Cauchy and thus converges: there exists a family $(Z_t^i)_{i \in S, t \geq 0}$ of càdlàg non-negative adapted processes such that for all $T \geq 0$,

$$\lim_n \sum_{i \in S} p_i E\left[\int_0^T |dZ_s^i - dZ_s^{i,n}|\right] = 0.$$

Passing to the limit the recursive definition of Z_t^n we deduce that the family of processes solves the expression of Definition 5.43. Finally, Lemma 5.47-(iii) implies that $\sup_n u_t^n < +\infty$ for all $t \geq 0$, from which $\sum_{i \in S} p_i E[Z_t^i] < +\infty$ as desired. $\qquad\square$

5.4.4. Galton-Watson formulation and likelihood expression

Now we will produce some quantitative estimates for a Hawkes process Z, based on the definition of an auxiliary Galton-Watson process W representing the branching of Z. For simplicity we will restrict to the case of a univariate linearly self-exciting counting process:

$$S - \{\lambda\}, \quad \mathcal{L} = (\lambda, \lambda), \quad h_x = \lambda_0 + x, \quad \phi_{xx} = \varphi_\mu$$

that is the density of a measure ν called infectivity measure. The intensity becomes in this case

$$\lambda(t, \omega) = \lambda_0 + \sum_{0 < t_i(\omega) < t} \varphi_\nu(t - t_i(\omega)) = \int_0^t \varphi_\nu(t - u)dZ_u.$$

Definition 5.49. A Galton-Watson process $(W_n)_{n \geq 0}$ with offspring distribution $p = (p_k)_{k \geq 0}$ is a discrete-time Markov chain taking values in the set $\mathbb{N}$ whose transition probabilities are as follows:

$$P\{W_{n+1} = k | W_n = h\} = (p^{*h})_k$$

Where p^{*h} denotes the hth convolution of the distribution p. The default initial state is $W_0 \equiv 1$.

The recursive structure leads to a simple set of relations among the probability generating functions of the random variables W_n:

Proposition 5.50. *Denote by $\eta_n(t) = E[t^{W_n}]$ the probability generating function of the random variable W_n, and by $\eta(t) = \sum_{k=0}^\infty p_k t^k$ the probability generating function of the offspring distribution. Then η_n is the n-fold composition of η by itself, that is*

$$\eta_0 = Id, \quad \eta_n = \eta^{\circ n}, \quad \forall n > 0.$$

Proof. We use the recursive structure directly to deduce that W_{n+1} is the sum of W_1 conditionally independent copies of W_n . Thus,

$$\eta_{n+1}(t) = E[t^{W_{n+1}}] = E[\eta_n^{W_1}] = \eta(\eta_n(t)).$$

By induction on n, this is the $(n+1)^{\text{th}}$ iterate of the function $\eta(t)$.   □

Corollary 5.51. *Let* $m := \sum_k^\infty k p_k < +\infty$ *be the mean offspring; then the expected size of the* n^{th} *generation is* $E[W_n] = m^n$. *Moreover if* $\sigma := Var[W_1]$ *is the variance of the offspring, then the variance of the size of the* n^{th} *generation is* $Var[W_n] = (\sigma + m^2 + m)\frac{1-m^n}{1-m}m^{n-1} + m^n - m^{2n}$.

Proof. From the properties of the generating functionals we have that

$$E[W_n] = \lim_{t \to 1^-} \eta_n'(t) = \lim_{t \to 1^-} \frac{d}{dt}\eta^{\circ n}(t) = m^n.$$

In the same way we can calculate the second statement:

$$Var[W_n] = E[W_n^2 - E[W_n]^2] = \lim_{t \to 1^-} \eta_n''(t) + \eta_n'(t) - \left(\eta_n'(t)\right)^2$$

$$= \lim_{t \to 1^-} \eta''(t)\eta'(t)^{n-1}\sum_{k=0}^{n-1} \eta'(t)^k + \eta'(t)^n - \eta'(t)^{2n}$$

$$= (E[W_1^2] - m)\frac{1-m^n}{1-m}m^{n-1} + m^n - m^{2n}$$

$$= (\sigma + m^2 + m)\frac{1-m^n}{1-m}m^{n-1} + m^n - m^{2n}$$

that is the wanted expression.   □

   Hence if $m < 1$ the limit size of the process is the finite sum of the geometric series $\sum_n^\infty m^n = \frac{1}{1-m}$ and the Galton-Watson process W_n is called subcritical, otherwise it tends to an infinite size and the process is called supercritical if $m > 1$, critical if $m = 1$.

   From a heuristic point of view, the points of a Hawkes process are of two types: 'immigrants' without ancestors in the process, and 'offspring' that are produced by existing points. The immigrants arrive according to a homogenous Poisson process with constant rate λ_0 while the offspring arise as elements of a NHPP that is associated with some point already constructed. Each immigrant has the potential to produce descendants whose numbers in successive generations constitute a Galton-Watson branching process.

Definition 5.52. Let Z be a Hawkes process with intensity $\lambda(t, \omega)$ defined above. Let Z^p be a NHPP with intensity equal to $\varphi_\nu(t)$. We define the related Galton-Watson process $(W_n^Z)_{n \geq 0}$, that has offspring distribution equal to the law of

$$\lim_{t \to +\infty} Z_t^p = \int_{\mathbb{R}_+} \varphi_\nu(t) dt = \nu(\mathbb{R}_+).$$

To be more accurate we can trivially decompose the related point process measure N to the sum of two random measures: N_0 ruled by the intensity λ_0 and representing the immigrants, and $N_c := N - N_0$ representing the offspring. The related counting process are Z^0 and Z^c respectively. Because of its definition it is straightforward to define the intensity function of the offspring conditional to the outcome $(t_i^{Z^0})_{i>0}$ of the process Z^0

$$\lambda^{Z^c}(t, \omega) = \sum_{0 < t_i^{Z^0}(\omega) < t} \varphi_\nu(t - t_i^{Z^0}(\omega)) + \sum_{0 < t_j^{Z^c}(\omega) < t} \varphi_\nu(t - t_j^{Z^c}(\omega));$$

using this conditional decomposition it is possible to define

$$\lambda^{Z^{c,i}}(t, \omega) := \varphi_\nu(t - t_i^{Z^0}(\omega)) + \sum_{0 < t_j^{Z^{c,i}}(\omega) < t} \varphi_\nu(t - t_j^{Z^{c,i}}(\omega)),$$

and it is easy to see that $\lambda^{Z^c}(t, \omega) = \sum_i \lambda^{Z^{c,i}}(t, \omega)$. Each of these conditional sub-processes $(Z^{c,i})_{i>0}$ represents the offspring of a single immigrant point and its random limit size $1 + \lim_{t \to +\infty} Z_t^{c,i}$ follows the law of the total size of the related Galton-Watson process W^Z. Passing to the point process measures formulation, we can prove in the same way that for each Borel set $A \in \mathcal{B}(\mathbb{R}_+)$ we have that $E[N(A)] \leq \frac{E[N^0(A)]}{1-m}$.

Proposition 5.53. *Let Z be a Hawkes process with $E[\int_{\mathbb{R}_+} \varphi_\nu(t) dt] = m < 1$. Then the process is almost surely finite, and for each $t \in \mathbb{R}_+$*

$$E[Z_t] \leq \frac{E[Z_t^0]}{1 - m}.$$

Proof. This inequality holds because the set $\mathbb{R}_+$ contains all the generations of the related branching process W. We have that

$$E[Z_t] = E[Z_t^0 + Z_t^c] = E[Z_t^0] + E\left[\sum_i Z_t^{c,i}\right],$$

but because of the independence of Z^0 and the family of identically distributed random variables $Z^{c,i}$, it is easy to obtain

$$= E[Z_t^0] + E[Z_t^0]E[Z_t^{c,1}] = E[Z_t^0](1 + E[Z_t^{c,1}]).$$

The expression

$$1 + E[Z_t^{c,1}] \leq 1 + E\left[\lim_{t \to +\infty} Z_t^{c,1}\right] = \frac{1}{1-m}$$

that is the limit size of the related Galton-Watson process. Hence we obtain the wanted bound.   □

Focusing on the parameters of an Hawkes process, we report the likelihood expression for an arbitrary time record. We define a univariate model, based on a linear Hawkes process with an exponential decay of self-exciting: we set

$$\varphi_v(t - s) = h \exp(-k(t - s))$$

where h, k with λ_0 are the three main parameters that drive the model. The multivariate case implemented in Chapter 4 corresponds to assuming

$$G = (\mathcal{S}, \mathcal{E}) = ((1, \dots, M), (j, j)_{j=1,\dots,M}),$$
$$h_j(x) = h(x) = q(j)\lambda_0 + x,$$
$$\varphi_{j,j} = \varphi_v,$$

where $q(j)$ is a long-term (base rate) spatial probability of vent opening in the subset j of the caldera partition in M zones (defined in Chapter 2). The time parameters have a simple interpretation with the help of the related Galton-Watson process, explained in Chapter 4:

$$\mu = h/k, \quad T = \ln(20)/k, \quad n = \lambda_0 T_0,$$

where μ is the mean offspring (the expected size of the first generation of a single element), T the 95th percentile of the infectivity measure (a threshold after whom only a 5% of the initial self-exciting remains), and n the mean number of base rate eruptions on the time domain $[0, T_0]$. We conclude reporting a result about the likelihood of a time record for a univariate Hawkes process, that it is easy to generalize for obtaining Proposition 4.5.

Proposition 5.54. *Let* $((t_i)_{i=1,\dots,n}, T_0 = t_{n+1})$ *be an increasing set of times, let* Z *be a Hawkes process with intensity function* λ, *then the likelihood of* $(t_i)_{i=1,\dots,n}$ *being the instants of the points of* Z *is obtained by*

$$L\left((t_i)_{i=1,\dots,n}, T_0\right) = \prod_{i=1}^{n} \lambda(t_i) \exp\left(-\int_{t_i}^{t_{i+1}} \lambda(s)ds\right)$$

$$= \left(\prod_{i=1}^{n} \lambda(t_i)\right) \exp\left(-\int_{0}^{T_0} \lambda(s)ds\right),$$

where

$$\int_{0}^{T_0} \lambda(s)ds = \lambda_0 T_0 - \sum_{i=1}^{n} \left(\frac{h}{k}\left(\exp(-k(T_0 - t_i)) - 1\right)\right),$$

and

$$\log L((t_i)_{i=1,\dots,n}, T_0) = \sum_{i=1}^{n} \log \lambda(t_i) - \lambda_0 T_0$$

$$+ \sum_{i=1}^{n} \frac{h}{k}\left(\exp(-k(T_0 - t_i)) - 1\right).$$

Chapter 6
Supporting information

6.1. Summary

In this chapter we include additional material concerning some technical details of the study. First it is reported the seed questionnaire that was adopted for obtaining the experts' scores (from W.P Aspinall, personal communication), the list of the true values is available on request. Then there is the complete and anonymous response list given by each expert to the seed and the target questions, including some range graphs of the results. After that are reported all the most important computer codes that were implemented during this study: all of them have been developed in the R software environment (*e.g.* [135]) and are referred as R-codes. At last there is a list of the principal symbols adopted in this study.

6.2. The expert judgement data on Campi Flegrei

6.2.1. Seed questionnaire

In the following it is reported a copy of the questionnaire that was adopted for obtaining the experts scores.

These are the 'seed' questions for calibrating individual expert's inputs, 'informativeness' and 'expected relative frequency (ERF)' in order to compute relative scores via EXCALIBUR and R software, to produce weightings for pooling responses in the elicitation of Logic Tree target items. Please provide both your 'credible range' of uncertainty (low value ↔ high value), and your 'central' estimate of the median value. The credible range should indicate the lowest and highest values you believe must encompass the 'true' answer with about 90% confidence (*i.e.* there is only a 5% chance the value falls below your lower value, and only a 5% chance it is higher than your upper value). For calibration and informativeness assessments your 'central' estimate should represent the median (50%ile) value of the uncertainty distribution - *i.e.* the value at which you judge there is an equal likelihood that the true realization will be above or below this value; instead for ERF calculation and Monte Carlo aggregation purposes the estimate is assumed as the mode, or most likely value. The distribution shape of your credible range need not be symmetric about the median.

1. Given your knowledge of the frequency of activity since the Neapolitan Yellow Tuff eruption $\sim$ 15 ka, what is the mean recurrence interval of eruptions at Campi Flegrei volcano, in years, in the last 15 ka (Data source: [53, 122])

low end value (5%ile)	median (50%ile)	high end value (95%ile)

2. At Campi Flegrei what percentage of eruptions in the last 15 ka has taken place from a vent located within a 4 km radius from the Temple of Serapis in Pozzuoli?

low end value (5%ile)	median (50%ile)	high end value (95%ile)

3. What percentage of explosive eruptions at Campi Flegrei volcano in the last 15 ka generated 10 cm isopachs that covered an area more than 500 km^2? (Source [122]). (As a percentage of the 22 deposits measured.)

low end value (5%ile)	median (50%ile)	high end value (95%ile)

4. A 350 m borehole has been drilled into the Colli Albani volcanic district. Gas and water samples were taken during a blow-out. From the water sample, what was the amount of HCO_3, in meq/l ?

low end value (5%ile)	median (50%ile)	high end value (95%ile)

5. For violent strombolian type eruptions at Vesuvius what is the mean area in km^2 enclosed by the 10 cm isopach (Data source: [6])?

low end value (5%ile)	median (50%ile)	high end value (95%ile)

6. At Vesuvius, what percentage of eruptions of that can be classified as 'violent strombolian' or larger in the last 20,000 years show evidence (either historical or geological) for Pyroclastic Density Currents (PDCs) with runouts > 2 km?

low end value (5%ile)   median (50%ile)   high end value (95%ile)

7. 1. What percentage of eruptions of sub-Plinian magnitude or larger at Vesuvius in the last 20,000 years have involved magmas of trachytic composition? (Data source: [35].)

low end value (5%ile)   median (50%ile)   high end value (95%ile)

8. At Mt Etna, what was the total area in km² covered by lava flows erupted during the 20th Century? Note this is not the actual proportion of the volcano that has been covered at any time but the total numerical area covered, when individual lava flow areas are summed (source [5]).

low end value (5%ile)   median (50%ile)   high end value (95%ile)

9. What is the volume of magma (DRE) in millions of m³ erupted as lava flows during the 20th Century at Etna volcano? (i.e. if you believe the median value is 100 × 106 m³, then enter 100 as your response in the median box)

low end value (5%ile)   median (50%ile)   high end value (95%ile)

10. In km², what is the area of the simplified structure of the outer Campi Flegrei caldera depression (e.g. as mapped by [2])?

low end value (5%ile)   median (50%ile)   high end value (95%ile)

11. [143] report the results of using the Multi-GAS technique to measure gas compositions in plumes at Etna in 2005 and 2006. They ascribe GOOD quality to four sets of measurements from the Voragine crater in which measurements of all three ratios: H_2O/CO_2, H_2O/SO_2 and $CO_2/SO - 2$ were obtained. For these four sets of measurements, if each pairing of H_2O/CO_2 ratio and H_2O/SO_2 ratio is combined to produce a counterpart pseudo-ratio estimate for CO_2/SO_2, then differences are found with respect to the measured CO_2/SO_2 ratio values. If each of these differences is expressed as a percentage of the corresponding measured CO_2/SO_2 ratio, then what is the overall range of these differences for all four cases? (*I.e.* what is the value of [highest numerical %age difference - lowest numerical %age difference] when, for example, given a highest value of +100% and a lowest value of −80%, respectively, would therefore represent a numerical range of 180).

low end value (5%ile)   median (50%ile)   high end value (95%ile)

12. In a recent work modelling the magma dynamics and collapse mechanisms during four well-known historic caldera-forming events, one model input was the time duration of magma evacuation before caldera block began to subside. Based on previously reported data, what value, in hours, was used for this duration in the case of Katmai 1912?

low end value (5%ile)   median (50%ile)   high end value (95%ile)

13. The same study quoted some values for the observed total downward surface displacement of the caldera block in certain caldera-forming eruptions. What was the reported downward displacement for Pinatubo 1991, in metres?

low end value (5%ile)   median (50%ile)   high end value (95%ile)

14. A tiltmeter network in the Campi Flegrei area has been level surveyed several times in recent years. One levelling point is within 1 km of a tiltmeter in a tunnel under Mount Olibano, Pozzuoli. What was the change in height of that levelling point from November 2004 to December 2006, in mm [+ for up, − for down]?

low end value (5%ile)   median (50%ile)   high end value (95%ile)

15. 1. Juvenile pumice clasts from a pyroclastic flow in the uppermost On-
    oda Formation (Chijimizawa-Tuff), Onikobe volcano, NE Japan, were
    classified by [113] into three colour types: white, gray, and dark gray.
    They determined experimentally the evolution of the vesicularity-per-
    meability relation during a single decompression of the white and gray
    pumices, and obtained a value for the critical vesicularity (Φ_C - ex-
    pressed as a percentage). This they compared with published power-
    law fits for Φ_C from data of various eruption products with different
    eruption styles from several different volcanoes. What is the ratio of
    the critical vesicularity for their Onoda white and gray pumices to the
    typical critical vesicularity value from the published power-law rela-
    tions? (*e.g.* if you think Φ_C pumices = 75% and if Φ_C power-law
    = 50%, then your ratio equals 1.5; if Φ_C pumices = 25% and if Φ_C
    power-law = 50%, the ratio equals 0.5).

low end value (5%ile)   median (50%ile)   high end value (95%ile)

16. Starting in 2004, a network of Sacks-Evertson borehole strainmeters
    (dilatometers) were installed in Campi Flegrei and near Vesuvius. At
    one site, near Toiano, close to the Campi Flegrei uplift centre, the tem-
    perature at instrument depth ($\sim$ 120 m) was high (60° C) on installa-
    tion, and found to be increasing steadily over some months, perhaps
    due to hydrothermal fluid flows. For the observed rate of increase,
    how long would it take for the hole temperature to increase by 1° C?
    Please state your time units (hours, days, weeks…years, decades).

low end value (5%ile)   median (50%ile)   high end value (95%ile)

We remark that questions 14 and 16 were excluded from the analysis
based on CM, because no one of the experts got values near the true
answers. This with ERF method was not needed. The cut-off on the
calibration score of CM was modulated for including at least 4 of the
experts. We did not average informativeness globally, but instead we

compared it separately for each question (item weights): this increased the DM score.

6.2.2. Experts responses list

In the sequel are reported all the expert responses to the seed and the target questions. Mover are also produced the range graphs for visualize their single uncertainty ranges on the seed questions. The true values are reported on the graphs with a # symbol. We remark that in the list have been also included the following 4 additional questions that we did not implement in the analysis because they were not additionally informative.

[Q7A] For defining the positional long-term probability of a new vent opening, and considering only the events of the III Epoch, what weights would you assign to a single vent which occurred before the AMS eruption? (11 vents identified, including AMS) (%)

[Q7B] For defining the positional long-term probability of a new vent opening, and considering only the events of the III Epoch, what weights would you assign to a single vent which occurred after the AMS eruption? (18 vents identified, including Monte Nuovo) (%)

[Q13A] What is the probability that the next Campi Flegrei eruption involves two (or more?) different (*i.e.* located a few kilometres from each other) vents being simultaneously active (*i.e.* within a few weeks of each other, or less)? (%)

[Q14A] Considering the areas that have been invaded by PDCs from Campi Flegrei in the three epochs and based on the information provided, what would be the typical linear percentage error on the radius of the invaded area (assumed as circular) in the underestimation of the PDC invasion areas (% of the radius)?

In particular Q7A and Q7B did not highlight a remarkable difference on the relative importance of the event location before and after AMS for what concern the production of the vent opening probability map. The responses to Q13A were not different from the responses to Q13 (see Table 2.1), which focused on the frequency of simultaneous events in the past eruptions instead that on the probability of the next events to be simultaneous. The question Q14A was redundant with Q14, with the difference that the first asked an estimate in percentage underestimation whereas the second an absolute underestimation (in meters); this second approach was preferred.

```
ELICITATION SESSION RESULTS
EXPERT 1 [CM=3.2% - ERF=11.9% - EW=12.5%]

SEED QUESTIONS

Exp1    1   CF_intervals log  1.50000E+0002  2.00000E+0002  2.60000E+0002
Exp1    2 %_vent_Serapis UNI  4.00000E+0001  6.00000E+0001  8.00000E+0001
Exp1    3      %isopachs UNI  9.90000E+0000  2.00000E+0001  4.00000E+0001
Exp1    4           HCO3 log  5.00000E-0002  5.00000E+0001  5.00000E+0003
Exp1    5 Ves_10cm_isoph UNI  3.00000E+0001  1.00000E+0002  3.00000E+0002
Exp1    6      %2km_PDCs UNI  1.50000E+0001  2.50000E+0001  4.00000E+0001
Exp1    7     %trachytic UNI  1.00000E+0001  2.00000E+0001  6.00000E+0001
Exp1    8 Area_lava_flow log  3.00000E+0002  5.00000E+0002  1.20000E+0003
Exp1    9     Vol_magma  UNI  1.00000E+0002  3.00000E+0002  8.00000E+0002
Exp1   10 Area_CFc_depre UNI  1.00000E+0002  1.20000E+0002  1.50000E+0002
Exp1   11 Gas_ratio_diff UNI  5.00000E+0001  2.00000E+0002  5.00000E+0002
Exp1   12 Magma_Evac_tim UNI  6.00000E+0000  2.40000E+0001  4.00000E+0001
Exp1   13     Surf._drop UNI  5.00000E+0000  2.00000E+0001  6.00000E+0001
Exp1   14  Height_change UNI -2.00000E+0001  3.00000E+0001  1.00000E+0002
Exp1   15     Crit_vesic UNI  3.00000E-0001  1.00000E+0000  1.70000E+0000
Exp1   16    Time_+1degC UNI  3.00000E-0002  3.00000E-0001  1.00000E+0000

EXPERT-WISE RANGE GRAPHS OF THE SEED QUESTIONS RESPONSES
  1(1)              [--------*-------]
  Real ::::::::::::::::::::::::#::::::::::::::::::::::::::::::::::::::::::::::::::

  2(U)                      [---------------*---------------]
  Real ::::::::::::::::::::::::#::::::::::::::::::::::::::::::::::::::::::::::::::

  3(U)     [------*--------------]
  Real ::::#:::::::::::::::::::::::::::::::::::::::::::::::::::::::::::::::::::::::

  4(1) [----------------------------------------*-----------------------]
  Real :::::::::::::::::::::::::::::::::::::::::::#::::::::::::::::::::::::::::::::

  5(U)     [-----------------*-----------------------------------------------]
  Real ::::::::#:::::::::::::::::::::::::::::::::::;;;;;;;;;;;;;;;'''''''''''''''''''iiii

  6(U)     [---------*-------------]
  Real :::::::::::::::::::::#::::::::::::::::::::::::::::::::::::::::::::::::::::::

  7(U) |-------*-----------------------------]
  Real :::::::::::::::::::::::::#:::::::::::::::::::::::::::::::::::::::::::::::::::

  8(1)                                       [-------*-------------]
  Real ::::::::::::::::::::::::::::::#::::::::::::::::::::::::::::::::::::::::::::::

  9(U) [*]
  Real ::::#:::::::::::::::::::::::::::::::::::::::::::::::::::::::::::::::::::::::

 10(U)          [----*-------]
  Real :::::::::::::::::::::::::::::::::::::::::::::#:::::::::::::::::::::::::::::::

 11(U)      [--------------------*---------------------------------------]
  Real :#::::::::::::::::::::::::::::::::::;;;;;;;;;;;;;;;;'''''''''''''::::::::::::::

 12(U)   [----------*---------]
  Real :::::::::::::::::::#:::::::::::::::::::::::::::::::::::::::::::::::::::::::::

 13(U) [-*-------]
  Real :::::::::::::::::::::::::::::::::::::::::::::::::::::::::::::::::::::::::::::#

 14(U) [-------------------------*-----------------------------------------]
  Real :::::::::::::::::::::::::::::::::#:::::::::::::::::::::::::::::::::::::::::::

 15(U)   [----*----]
  Real :::::::::::::::#:::::::::::::::::::::::::::::::::::::::::::::::::::::::::::::

 16(U) |
  Real :::::::::::::::::::::::::::::::::::::::::#::::::::::::::::::::::::::::::::::

TARGET QUESTIONS
Exp1   17 Q1Wt_5variable uni  7.00000E+0001  9.00000E+0001  9.50000E+0001
Exp1   18   Q2Wt_Uniform UNI  5.00000E+0000  1.00000E+0001  3.00000E+0001
Exp1   19 Q3Wt_vents_loc UNI  5.00000E+0001  7.00000E+0001  9.00000E+0001
Exp1   20 Q4Wt_structure UNI  1.00000E+0001  3.00000E+0001  5.00000E+0001
Exp1   21   Q5Wt_I_vent  UNI  5.00000E+0000  2.00000E+0001  4.00000E+0001
```

```
Exp1   22    Q6Wt_II_vent UNI  1.00000E+0001  3.00000E+0001  5.00000E+0001
Exp1   23   Q7Wt_III_vent UNI  3.00000E+0001  5.00000E+0001  7.00000E+0001
Exp1   24     Q7a_pre-AMS UNI  3.00000E+0001  4.50000E+0001  6.00000E+0001
Exp1   25     Q7b_postAMS UNI  4.00000E+0001  5.50000E+0001  7.00000E+0001
Exp1   26     Q8Wt_faults UNI  3.00000E+0001  6.00000E+0001  8.00000E+0001
Exp1   27  Q9Wt_fractures UNI  2.00000E+0001  4.00000E+0001  7.00000E+0001
Exp1   28    Q10_missed_I UNI  5.00000E+0000  7.00000E+0000  1.50000E+0001
Exp1   29   Q11_missed_II UNI  1.00000E+0000  2.00000E+0000  3.00000E+0000
Exp1   30  Q12_missed_III UNI  1.00000E+0000  2.00000E+0000  4.00000E+0000
Exp1   31  Q13%_Multiples UNI  5.00000E+0000  1.00000E+0001  3.00000E+0001
Exp1   32  Q13a_PnextMult UNI  1.00000E+0000  1.00000E+0001  3.00000E+0001
Exp1   33  Q14a_SpatUncer uni  1.50000E+0002  3.00000E+0002  1.00000E+0003
Exp1   34  Q14RadiusError uni  5.00000E+0000  1.00000E+0001  2.50000E+0001
```

ELICITATION SESSION RESULTS
EXPERT 2 [CM=0.0% - ERF=7.8% - EW=12.5%]

```
SEED QUESTIONS .
Exp2    1     CF_intervals log  1.75000E+0002  2.00000E+0002  2.25000E+0002
Exp2    2  %_vent_Serapis UNI  5.00000E+0001  6.00000E+0001  7.00000E+0001
Exp2    3       %isopachs UNI  5.00000E+0000  1.00000E+0001  3.00000E+0001
Exp2    4            HCO3 log  1.00000E+0000  1.00000E+0001  1.00000E+0002
Exp2    5  Ves_10cm_isoph UNI  1.00000E+0002  2.00000E+0002  3.00000E+0002
Exp2    6       %2km_PDCs UNI  4.00000E+0001  5.00000E+0001  7.00000E+0001
Exp2    7       %trachytic UNI  1.50000E+0001  3.00000E+0001  5.00000E+0001
Exp2    8  Area_lava_flow log  2.00000E+0001  2.50000E+0001  4.00000E+0001
Exp2    9       Vol_magma UNI  5.00000E-0002  1.00000E+0000  3.00000E+0000
Exp2   10  Area_CFc_depre UNI  1.00000E+0002  1.50000E+0002  2.00000E+0002
Exp2   11  Gas_ratio_diff UNI  3.00000E+0001  4.00000E+0001  1.00000E+0002
Exp2   12  Magma_Evac_tim UNI  2.40000E+0001  4.80000E+0001  1.20000E+0002
Exp2   13      Surf._drop UNI  1.00000E+0002  1.50000E+0002  2.00000E+0002
Exp2   14   Height_change UNI  5.00000E+0000  1.50000E+0001  2.50000E+0001
Exp2   15       Crit_vesic UNI  5.00000E-0001  1.00000E+0000  1.25000E+0000
Exp2   16      Time_+1degC UNI  1.00000E-0002  3.30000E-0001  1.00000E+0000
```

```
EXPERT-WISE RANGE GRAPHS OF THE SEED QUESTIONS RESPONSES
 1(1)                   [---*---]
Real    ::::::::::::::::::::::::#:::::::::::::::::::::::::::::::::::::::::::::

 2(U)                             [-------*-------]
Real    :::::::::::::::::::::::#::::::::::::::::::::::::::::::::::::::::::::::

 3(U) [---*--------------]
Real    ::::#:::::::::::::::::::::::::::::::::::::::::::::::::::::::::::::::::

 4(1)                [-----------*-----------]
Real    ::::::::::::::::::::::::::::::::::::#:::::::::::::::::::::::::::::::::

 5(U)                  [------------------------*-----------------------]
Real    ::::::::::#:::::::::::::::::::::::::::::::::::::::::::::::::::::::::::

 6(U)                      [---------*------------------]
Real    :::::::::::::::::::::#::::::::::::::::::::::::::::::::::::::::::::::::

 7(U)      [------------*-----------------]
Real    ::::::::::::::::::::::#:::::::::::::::::::::::::::::::::::::::::::::::

 8(1)      [---*-------]
Real    ::::::::::::::::::::::::::::#:::::::::::::::::::::::::::::::::::::::::

 9(U)  |
Real    ::::#:::::::::::::::::::::::::::::::::::::::::::::::::::::::::::::::::

 10(U)       [------------*-----------]
Real    ::::::::::::::::::::::::::::::::::::::::::::#:::::::::::::::::::::::::

 11(U)     [*--------]
Real    :#::::::::::::::::::::::::::::::::::::::::::::::::::::::::::::::::::::

 12(U)            [--------------*------------------------------------------]
Real    ::::::::::::::#:::::::::::::::::::::::::::::::::::::::::::::::::::::::

 13(U)                 [--------*--------]
Real    :::::::::::::::::::::::::::::::::::::::::::::::::::::::::::::::::::::#
```

```
   14(U)                [-----*-----]
   Real   ::::::::::::::::::::::::::::::::::#::::::::::::::::::::::::::::::::::::

   15(U)   [---*-]
   Real   :::::::::::::::#::::::::::::::::::::::::::::::::::::::::::::::::::::::::

   16(U)  |
   Real   ::::::::::::::::::::::::::::::::::::#:::::::::::::::::::::::::::::::::::

   TARGET QUESTIONS
   Exp2   17 Q1Wt_5variable uni  8.50000E+0001  9.00000E+0001  9.50000E+0001
   Exp2   18    Q2Wt_Uniform UNI  5.00000E+0000  1.00000E+0001  1.50000E+0001
   Exp2   19 Q3Wt_vents_loc UNI  5.50000E+0001  6.50000E+0001  8.00000E+0001
   Exp2   20 Q4Wt_structure UNI  2.50000E+0001  3.50000E+0001  4.50000E+0001
   Exp2   21    Q5Wt_I_vent UNI  2.00000E+0001  2.50000E+0001  4.00000E+0001
   Exp2   22   Q6Wt_II_vent UNI  3.00000E+0001  3.50000E+0001  4.50000E+0001
   Exp2   23  Q7Wt_III_vent UNI  3.00000E+0001  4.00000E+0001  5.00000E+0001
   Exp2   24    Q7a_pre-AMS UNI  3.00000E+0001  4.00000E+0001  5.00000E+0001
   Exp2   25    Q7b_postAMS UNI  5.00000E+0001  6.00000E+0001  7.00000E+0001
   Exp2   26    Q8Wt_faults UNI  4.00000E+0001  6.00000E+0001  7.00000E+0001
   Exp2   27 Q9Wt_fractures UNI  3.00000E+0001  4.00000E+0001  5.00000E+0001
   Exp2   28    Q10_missed_I UNI  5.00000E+0000  6.00000E+0000  9.00000E+0000
   Exp2   29   Q11_missed_II UNI  0.00000E+0000  1.00000E+0000  2.00000E+0000
   Exp2   30  Q12_missed_III UNI  1.00000E+0000  2.00000E+0000  3.00000E+0000
   Exp2   31  Q13%_Multiples UNI  5.00000E+0000  1.50000E+0001  3.00000E+0001
   Exp2   32 Q13a_PnextMult UNI  5.00000E+0000  1.50000E+0001  3.00000E+0001
   Exp2   33 Q14a_SpatUncer uni  1.00000E+0002  4.00000E+0002  8.00000E+0002
   Exp2   34 Q14RadiusError uni  5.00000E+0000  1.50000E+0001  2.50000E+0001


   ELICITATION SESSION RESULTS
   EXPERT 3 [CM=0.0% - ERF=7.3% - EW=12.5%]

   SEED QUESTIONS
   Exp3    1    CF_intervals log  2.20000E+0002  2.50000E+0002  2.70000E+0002
   Exp3    2 %_vent_Serapis UNI  4.00000E+0000  8.00000E+0000  1.30000E+0001
   Exp3    3      %isopachs UNI  9.90000E+0000  1.80000E+0001  3.00000E+0001
   Exp3    4          IIΔp0 log  5.00000E10000  2.00000E10003  0.50000E10003
   Exp3    5 Ves_10cm_isoph UNI  9.00000E+0001  1.50000E+0002  2.00000E+0002
   Exp3    6       %2km_PDCs UNI  6.00000E+0001  8.00000E+0001  8.50000E+0001
   Exp3    7     %trachytic UNI  6.00000E+0001  7.00000E+0001  9.00000E+0001
   Exp3    8 Area_lava_flow log  1.50000E+0001  4.00000E+0001  6.00000E+0001
   Exp3    9     Vol_magma UNI  1.00000E+0000  3.00000E+0000  5.00000E+0000
   Exp3   10 Area_CFc_depre UNI  1.90000E+0002  2.00000E+0002  2.10000E+0002
   Exp3   11 Gas_ratio_diff UNI  2.50000E+0001  1.00000E+0002  2.00000E+0002
   Exp3   12 Magma_Evac_tim UNI  3.00000E+0001  5.00000E+0001  1.00000E+0002
   Exp3   13    Surf._drop UNI  1.00000E+0002  1.80000E+0002  2.00000E+0002
   Exp3   14  Height_change UNI  2.00000E+0001  5.00000E+0001  8.00000E+0001
   Exp3   15     Crit_vesic UNI  5.00000E-0001  2.00000E+0000  3.00000E+0000
   Exp3   16    Time_+1degC UNI  2.00000E-0002  8.00000E-0002  5.00000E-0001


   EXPERT-WISE RANGE GRAPHS OF THE SEED QUESTIONS RESPONSES
    1(l)                            [---*--]
   Real   ::::::::::::::::::::::::::::#::::::::::::::::::::::::::::::::::::::::::

    2(U) [--*---]
   Real   ::::::::::::::::::::::::::::#::::::::::::::::::::::::::::::::::::::::::

    3(U)     [-----*--------]
   Real   ::::#:::::::::::::::::::::::::::::::::::::::::::::::::::::::::::::::::::

    4(l)                             [-------*]
   Real   ::::::::::::::::::::::::::::::::::::#:::::::::::::::::::::::::::::::::::

    5(U)             [--------------*------------]
   Real   :::::::::#::::::::::::::::::::::::::::::::::::::::::::::::::::::::::::::

    6(U)                            [--------------------*---]
   Real   ::::::::::::::::::#:::::::::::::::::::::::::::::::::::::::::::::::::::::

    7(U)                          [--------*-----------------]
   Real   ::::::::::::::::::#:::::::::::::::::::::::::::::::::::::::::::::::::::::

    8(l) [---------------*-----]
   Real   :::::::::::::::::::::::::::::::::#::::::::::::::::::::::::::::::::::::::
```

```
 9(U)  |
Real   ::::#:::::::::::::::::::::::::::::::::::::::::::::::::::::::::::::::::::

10(U)                                   [-*--]
Real   :::::::::::::::::::::::::::::::::::::::::::::::::::#:::::::::::::::::::::

11(U)     [----------*-------------]
Real   :#:::::::::::::::::::::::::::::::::::::::::::::::::::::::::::::::::::::::

12(U)                    [-----------*----------------------------]
Real   :::::::::::::::::#:::::::::::::::::::::::::::::::::::::::::::::::::::::::

13(U)                  [-------------*---]
Real   :::::::::::::::::::::::::::::::::::::::::::::::::::::::::::::::::::::::::#

14(U)                      [-----------------*-----------------]
Real   ::::::::::::::::::::::::::#:::::::::::::::::::::::::::::::::::::::::::::::

15(U)     [----------*-------]
Real   :::::::::::::#:::::::::::::::::::::::::::::::::::::::::::::::::::::::::::

16(U)  |
Real   :::::::::::::::::::::::::::::::::::#:::::::::::::::::::::::::::::::::::::
```

TARGET QUESTIONS

Exp3	17	Q1Wt_5variable	uni	7.50000E+0001	8.50000E+0001	9.50000E+0001
Exp3	18	Q2Wt_Uniform	UNI	5.00000E+0000	1.50000E+0001	2.00000E+0001
Exp3	19	Q3Wt_vents_loc	UNI	5.50000E+0001	6.00000E+0001	7.00000E+0001
Exp3	20	Q4Wt_structure	UNI	3.50000E+0001	4.00000E+0001	5.00000E+0001
Exp3	21	Q5Wt_I_vent	UNI	2.50000E+0001	3.00000E+0001	3.50000E+0001
Exp3	22	Q6Wt_II_vent	UNI	2.50000E+0001	3.00000E+0001	3.50000E+0001
Exp3	23	Q7Wt_III_vent	UNI	3.00000E+0001	4.00000E+0001	6.00000E+0001
Exp3	24	Q7a_pre-AMS	UNI	2.50000E+0001	3.00000E+0001	4.50000E+0001
Exp3	25	Q7b_postAMS	UNI	6.00000E+0001	7.00000E+0001	7.50000E+0001
Exp3	26	Q8Wt_faults	UNI	4.00000E+0001	4.50000E+0001	6.50000E+0001
Exp3	27	Q9Wt_fractures	UNI	5.00000E+0001	5.50000E+0001	6.50000E+0001
Exp3	28	Q10_missed_I	UNI	5.00000E+0000	7.00000E+0000	9.00000E+0000
Exp3	29	Q11_missed_II	UNI	0.00000E+0000	1.00000E+0000	2.00000E+0000
Exp3	30	Q12_missed_III	UNI	0.00000E+0000	1.00000E+0000	2.00000E+0000
Exp3	31	Q13%_Multiples	UNI	3.00000E+0000	5.00000E+0000	1.00000E+0001
Exp3	32	Q13a_PnextMult	UNI	2.00000E+0000	5.00000E+0000	1.00000E+0001
Exp3	33	Q14a_SpatUncer	uni	1.00000E+0002	3.00000E+0002	1.50000E+0003
Exp3	34	Q14RadiusError	uni	5.00000E+0000	1.00000E+0001	2.00000E+0001

ELICITATION SESSION RESULTS
EXPERT 4 [CM=0.0% - ERF=8.8% - EW=12.5%]

SEED QUESTIONS

Exp4	1	CF_intervals	log	3.00000E+0002	6.00000E+0002	1.00000E+0003
Exp4	2	%_vent_Serapis	UNI	4.00000E+0001	7.50000E+0001	9.50000E+0001
Exp4	3	%isopachs	UNI	2.00000E+0001	3.00000E+0001	5.00000E+0001
Exp4	4	HCO3	log	5.00000E+0002	3.00000E+0004	5.00000E+0004
Exp4	5	Ves_10cm_isoph	UNI	1.00000E+0002	1.50000E+0002	3.00000E+0002
Exp4	6	%2km_PDCs	UNI	1.50000E+0001	2.00000E+0001	2.50000E+0001
Exp4	7	%trachytic	UNI	1.50000E+0001	2.00000E+0001	3.00000E+0001
Exp4	8	Area_lava_flow	log	1.50000E+0002	2.50000E+0002	5.00000E+0002
Exp4	9	Vol_magma	UNI	3.00000E+0002	6.00000E+0002	1.20500E+0003
Exp4	10	Area_CFc_depre	UNI	1.20000E+0002	2.00000E+0002	3.50000E+0002
Exp4	11	Gas_ratio_diff	UNI	2.00000E+0001	1.10000E+0002	1.70000E+0002
Exp4	12	Magma_Evac_tim	UNI	4.00000E+0000	1.50000E+0001	2.80000E+0001
Exp4	13	Surf._drop	UNI	5.00000E+0001	1.00000E+0002	3.00000E+0002
Exp4	14	Height_change	UNI	-1.00000E+0001	2.00000E+0001	5.00000E+0001
Exp4	15	Crit_vesic	UNI	5.00000E-0001	1.50000E+0000	3.00000E+0000
Exp4	16	Time_+1degC	UNI	3.00000E-0002	1.50000E-0001	3.00000E-0001

EXPERT-WISE RANGE GRAPHS OF THE SEED QUESTIONS RESPONSES

```
 1(1)                              [---------------------*--------------]
Real   ::::::::::::::::::::::::#:::::::::::::::::::::::::::::::::::::::::::::::

 2(U)                          [---------------------------*--------------]
Real   :::::::::::::::::::::#:::::::::::::::::::::::::::::::::::::::::::::::::::

 3(U)         [-------*--------------]
Real   ::::#:::::::::::::::::::::::::::::::::::::::::::::::::::::::::::::::::::
```

```
   4(1)                                                   [---------------------*-]
   Real   :::::::::::::::::::::::::::::::::::::::::::#:::::::::::::::::::::::::::::::::::::

   5(U)                      [-----------*------------------------------------]
   Real   :::::::::#:::::::::::::::::::::::::::::::::::::::::::::::::::::::::::::::::::::::

   6(U)      [----*----]
   Real   :::::::::::::::::::::::#:::::::::::::::::::::::::::::::::::::::::::::::::::::::::

   7(U)      [---*--------]
   Real   :::::::::::::::::::::::#:::::::::::::::::::::::::::::::::::::::::::::::::::::::::

   8(1)                                [--------*----------]
   Real   :::::::::::::::::::::::::::::::::#:::::::::::::::::::::::::::::::::::::::::::::::

   9(U)   [*-]
   Real   ::::#::::::::::::::::::::::::::::::::::::::::::::::::::::::::::::::::::::::::::::

  10(U)                  [-------------------*------------------------------]
  Real   :::::::::::::::::::::::::::::::::::::::::::::::#:::::::::::::::::::::::::::::::::

  11(U)   [------------*--------]
  Real   :#::::::::::::::::::::::::::::::::::::::::::::::::::::::::::::::::::::::::::::::::

  12(U) [------*-------]
  Real   ::::::::::::::::#:::::::::::::::::::::::::::::::::::::::::::::::::::::::::::::::::

  13(U)            [--------*----------------------------------]
  Real   :::::::::::::::::::::::::::::::::::::::::::::::::::::::::::::::::::::::::::::::::#

  14(U)       [----------------*-----------------]
  Real   :::::::::::::::::::::::::::::::::::#:::::::::::::::::::::::::::::::::::::::::::::

  15(U)   [-------*----------]
  Real   ::::::::::::::::::#:::::::::::::::::::::::::::::::::::::::::::::::::::::::::::::::

  16(U) |
  Real   :::::::::::::::::::::::::::::::::::::::#:::::::::::::::::::::::::::::::::::::::::
```

```
TARGET QUESTIONS
Exp4   17 Q1Wt_5variable uni   8.00000E+0001   9.00000E+0001   9.50000E+0001
Exp4   18    Q2Wt Uniform UNI   5.00000E+0000   1.00000E+0001   2.00000E+0001
Exp4   19 Q3Wt_vents_loc UNI   5.50000E+0001   7.50000E+0001   8.50000E+0001
Exp4   20 Q4Wt_structure UNI   1.50000E+0001   2.50000E+0001   4.50000E+0001
Exp4   21    Q5Wt_I_vent UNI   1.00000E+0001   2.00000E+0001   3.00000E+0001
Exp4   22  Q6Wt_II_vent UNI   2.00000E+0001   3.50000E+0001   5.00000E+0001
Exp4   23 Q7Wt_III_vent UNI   3.00000E+0001   4.50000E+0001   7.50000E+0001
Exp4   24   Q7a_pre-AMS UNI   3.00000E+0001   4.00000E+0001   5.50000E+0001
Exp4   25   Q7b_postAMS UNI   4.50000E+0001   6.00000E+0001   7.00000E+0001
Exp4   26   Q8Wt_faults UNI   4.00000E+0001   6.00000E+0001   7.00000E+0001
Exp4   27 Q9Wt_fractures UNI   3.00000E+0001   4.00000E+0001   6.00000E+0001
Exp4   28   Q10_missed_I UNI   4.00000E+0000   6.00000E+0000   1.10000E+0001
Exp4   29  Q11_missed_II UNI   0.00000E+0000   1.00000E+0000   2.00000E+0000
Exp4   30 Q12_missed_III UNI   1.00000E+0000   2.00000E+0000   3.00000E+0000
Exp4   31 Q13%_Multiples UNI   5.00000E+0000   1.00000E+0001   3.00000E+0001
Exp4   32 Q13a_PnextMult UNI   5.00000E+0000   1.00000E+0001   3.00000E+0001
Exp4   33 Q14a_SpatUncer uni   1.50000E+0002   3.00000E+0002   7.50000E+0002
Exp4   34 Q14RadiusError uni   5.00000E+0000   1.00000E+0001   2.50000E+0001
```

ELICITATION SESSION RESULTS
EXPERT 5 [CM=6.8% - ERF=19.4% - EW=12.5%]

```
SEED QUESTIONS
Exp5    1    CF_intervals log   1.30000E+0002   2.10000E+0002   2.50000E+0002
Exp5    2 %_vent_Serapis UNI   3.00000E+0001   5.00000E+0001   6.00000E+0001
Exp5    3      %isopachs UNI   6.00000E+0001   7.50000E+0001   9.00000E+0001
Exp5    4           HCO3 log   2.00000E+0001   1.00000E+0002   3.00000E+0002
Exp5    5 Ves_10cm_isoph UNI   3.00000E+0001   5.00000E+0001   1.50000E+0002
Exp5    6      %2km_PDCs UNI   1.00000E+0001   3.00000E+0001   6.00000E+0001
Exp5    7     %trachytic UNI   4.00000E+0001   5.00000E+0001   7.00000E+0001
Exp5    8 Area_lava_flow log   1.00000E+0002   3.00000E+0002   7.00000E+0002
Exp5    9     Vol_magma UNI   5.00000E+0003   1.00000E+0004   1.30000E+0004
Exp5   10 Area_CFc_depre UNI   1.20000E+0002   1.30000E+0002   2.50000E+0002
Exp5   11 Gas_ratio_diff UNI   2.00000E+0001   3.00000E+0001   5.00000E+0001
Exp5   12 Magma_Evac_tim UNI   1.80000E+0001   2.40000E+0001   3.60000E+0001
Exp5   13     Surf._drop UNI   5.00000E+0001   2.00000E+0002   3.00000E+0002
```

```
Exp5   14   Height_change UNI  2.00000E+0001   3.00000E+0001   5.00000E+0001
Exp5   15     Crit_vesic UNI   1.00000E-0001   1.00000E+0000   5.00000E+0000
Exp5   16    Time_+1degC UNI   5.00000E-0001   1.00000E+0000   3.00000E+0000


EXPERT-WISE RANGE GRAPHS OF THE SEED QUESTIONS RESPONSES
  1(l)           [--------------*----]
Real  :::::::::::::::::::::::::::#:::::::::::::::::::::::::::::::::::::::::::::::
  2(U)                    [---------------*-------]
Real  :::::::::::::::::::::::::#:::::::::::::::::::::::::::::::::::::::::::::::::
  3(U)                                          [-----------*----------]
Real  ::::#:::::::::::::::::::::::::::::::::::::::::::::::::::::::::::::::::::::::
  4(l)                        [-------*-----]
Real  :::::::::::::::::::::::::::::::#:::::::::::::::::::::::::::::::::::::::::::::
  5(U)      [----*-----------------------]
Real  ::::::::#:::::::::::::::::::::::::::::::::::::::::::::::::::::::::::::::::::::
  6(U) [-----------------*------------------------]
Real  ::::::::::::::::::#:::::::::::::::::::::::::::::::::::::::::::::::::::::::::::
  7(U)             [--------*-----------------]
Real  :::::::::::::::::::::::::#:::::::::::::::::::::::::::::::::::::::::::::::::::
  8(l)             [------------------*-------------]
Real  :::::::::::::::::::::::::::::::#::::::::::::::::::::::::::::::::::::::::::::::
  9(U)          [-----------------*----------]
Real  ::::#:::::::::::::::::::::::::::::::::::::::::::::::::::::::::::::::::::::::::
 10(U)          [--*----------------------------]
Real  ::::::::::::::::::::::::::::::::::::::::::#:::::::::::::::::::::::::::::::::::
 11(U)     [-*-]
Real  :#::::::::::::::::::::::::::::::::::::::::::::::::::::::::::::::::::::::::::::
 12(U)           [--*-------]
Real  ::::::::::::::::::::::::::#::::::::::::::::::::::::::::::::::::::::::::::::::::
 13(U)         [------------------------*-----------------]
Real  :::::::::::::::::::::::::::::::::::::::::::::::::::::::::::::::::::::::::::::#
 14(U)                  [-----*-----------]
Real  :::::::::::::::::::::::::::::::::::#:::::::::::::::::::::::::::::::::::::::::::
 15(U) [-----*----------------------]
Real  ::::::::::::::::#:::::::::::::::::::::::::::::::::::::::::::::::::::::::::::::
 16(U) *]
Real  ::::::::::::::::::::::::::::::::::::#:::::::::::::::::::::::::::::::::::::::::

TARGET QUESTIONS
Exp5   17  Q1Wt_5variable uni   6.50000E+0001   8.00000E+0001   9.50000E+0001
Exp5   18     Q2Wt_Uniform UNI  5.00000E+0000   2.00000E+0001   3.50000E+0001
Exp5   19  Q3Wt_vents_loc UNI   5.00000E+0001   6.50000E+0001   8.00000E+0001
Exp5   20   Q4Wt_structure UNI  2.00000E+0001   3.50000E+0001   5.00000E+0001
Exp5   21     Q5Wt_I_vent UNI   1.00000E+0001   2.50000E+0001   4.00000E+0001
Exp5   22    Q6Wt_II_vent UNI   2.00000E+0001   5.00000E+0001   6.00000E+0001
Exp5   23   Q7Wt_III_vent UNI   2.00000E+0001   3.50000E+0001   6.00000E+0001
Exp5   24     Q7a_pre-AMS UNI   2.00000E+0001   4.00000E+0001   6.00000E+0001
Exp5   25    Q7b_postAMS UNI    4.00000E+0001   6.00000E+0001   8.00000E+0001
Exp5   26    Q8Wt_faults UNI    4.00000E+0001   6.00000E+0001   8.00000E+0001
Exp5   27  Q9Wt_fractures UNI   2.00000E+0001   4.00000E+0001   6.00000E+0001
Exp5   28     Q10_missed_I UNI  4.00000E+0001   7.00000E+0000   1.00000E+0001
Exp5   29   Q11_missed_II UNI   0.00000E+0000   1.00000E+0000   2.00000E+0000
Exp5   30  Q12_missed_III UNI   1.00000E+0000   2.00000E+0000   4.00000E+0000
Exp5   31   Q13%_Multiples UNI  5.00000E+0000   1.00000E+0001   2.00000E+0001
Exp5   32   Q13a_PnextMult UNI  5.00000E+0000   1.00000E+0001   2.00000E+0001
Exp5   33  Q14a_SpatUncer uni   1.00000E+0002   5.00000E+0002   1.00000E+0003
Exp5   34  Q14RadiusError uni   5.00000E+0000   1.50000E+0001   2.00000E+0001
```

ELICITATION SESSION RESULTS
EXPERT 6 [CM=0.0% - ERF=13.3% - EW=12.5%]

```
SEED QUESTIONS
Exp6    1    CF_intervals log  2.00000E+0002  2.50000E+0002  3.00000E+0002
Exp6    2 %_vent_Serapis UNI  7.00000E+0001  9.00000E+0001  9.50000E+0001
Exp6    3      %isopachs UNI  4.50000E+0000  9.00000E+0000  1.35000E+0002
Exp6    4          HCO3 log  5.00000E+0000  5.00000E+0001  1.00000E+0002
Exp6    5 Ves_10cm_isoph UNI  5.00000E+0001  1.00000E+0002  2.00000E+0002
Exp6    6       %2km_PDCs UNI  4.00000E+0001  6.00000E+0001  7.00000E+0001
Exp6    7     %trachytic UNI  5.00000E+0001  8.00000E+0001  9.00000E+0001
Exp6    8 Area_lava_flow log  2.00000E+0001  8.00000E+0001  1.50000E+0002
Exp6    9     Vol_magma UNI  2.00000E+0000  8.00000E+0001  1.50000E+0001
Exp6   10 Area_CFc_depre UNI  1.10000E+0002  1.30000E+0002  1.45000E+0002
Exp6   11 Gas_ratio_diff UNI  1.00000E+0001  5.00000E+0001  1.00000E+0002
Exp6   12 Magma_Evac_tim UNI  1.50000E+0001  3.00000E+0001  5.00000E+0001
Exp6   13    Surf._drop UNI  3.00000E+0001  5.00000E+0001  1.00000E+0002
Exp6   14 Height_change UNI  5.00000E+0000  1.50000E+0001  3.00000E+0001
Exp6   15    Crit_vesic UNI  1.00000E-0001  5.00000E-0001  1.50000E+0000
Exp6   16    Time_+1degC UNI  2.40000E+0000  4.80000E+0000  1.20000E+0001
```

```
EXPERT-WISE RANGE GRAPHS OF THE SEED QUESTIONS RESPONSES

  1(l)                        [------*-----]
Real   :::::::::::::::::::::::::::#:::::::::::::::::::::::::::::::::::::::::::::::

  2(U)                                         [---------------*--]
Real   :::::::::::::::::::::::::#:::::::::::::::::::::::::::::::::::::::::::::::::

  3(U) [--*--]
Real   ::::#:::::::::::::::::::::::::::::::::::::::::::::::::::::::::::::::::::::::

  4(l)                     [-----------*---]
Real   :::::::::::::::::::::::::::::::::::::#:::::::::::::::::::::::::::::::::::::

  5(U)          [-------------*------------------------]
Real   :::::::::#:::::::::::::::::::::::::::::::::::::::::::::::::::::::::::::::::

  6(U)                   [-------------------*---------]
Real   :::::::::::::::::::#:::::::::::::::::::::::::::::::::::::::::::::::::::::::

  7(U)                      [-------------------------*--------]
Real   :::::::::::::::::#::::::::::::::::::::::::::::::::::::::::::::::::::::::::::

  8(l)        [----------------------*---------]
Real   :::::::::::::::::::::::::#::::::::::::::::::::::::::::::::::::::::::::::::::

  9(U) |
Real   ::::#:::::::::::::::::::::::::::::::::::::::::::::::::::::::::::::::::::::::

 10(U)              [----*---]
Real   ::::::::::::::::::::::::::::::::::::::::#:::::::::::::::::::::::::::::::::::

 11(U) [----*-------]
Real   :#::::::::::::::::::::::::::::::::::::::::::::::::::::::::::::::::::::::::::

 12(U)        [--------*-----------]
Real   :::::::::::::::#:::::::::::::::::::::::::::::::::::::::::::::::::::::::::::::

 13(U)     [---*--------]
Real   :::::::::::::::::::::::::::::::::::::::::::::::::::::::::::::::::::::::::::::#

 14(U)               [-----*--------]
Real   :::::::::::::::::::::::::::::::::::::#::::::::::::::::::::::::::::::::::::::

 15(U) [-*-------]
Real   ::::::::::::::::#:::::::::::::::::::::::::::::::::::::::::::::::::::::::::::

 16(U) [*----]
Real   :::::::::::::::::::::::::::::::#:::::::::::::::::::::::::::::::::::::::::::::
```

```
TARGET QUESTIONS
Exp6   17 Q1Wt_5variable uni  6.50000E+0001  7.00000E+0001  8.00000E+0001
Exp6   18   Q2Wt_Uniform UNI  2.00000E+0001  3.00000E+0001  3.50000E+0001
Exp6   19 Q3Wt_vents_loc UNI  4.50000E+0001  5.00000E+0001  5.50000E+0001
Exp6   20 Q4Wt_structure UNI  4.50000E+0001  5.00000E+0001  5.50000E+0001
```

```
Exp6  21    Q5Wt_I_vent UNI  1.50000E+0001   2.00000E+0001   2.50000E+0001
Exp6  22    Q6Wt_II_vent UNI 1.50000E+0001   2.00000E+0001   2.50000E+0001
Exp6  23    Q7Wt_III_vent UNI 5.50000E+0001  6.00000E+0001   7.00000E+0001
Exp6  24    Q7a_pre-AMS UNI  4.50000E+0001   5.00000E+0001   5.50000E+0001
Exp6  25    Q7b_postAMS UNI  4.50000E+0001   5.00000E+0001   5.50000E+0001
Exp6  26    Q8Wt_faults UNI  6.00000E+0001   6.50000E+0001   7.00000E+0001
Exp6  27 Q9Wt_fractures UNI  3.00000E+0001   3.50000E+0001   4.00000E+0001
Exp6  28   Q10_missed_I UNI  5.00000E+0000   6.00000E+0000   7.00000E+0000
Exp6  29  Q11_missed_II UNI  1.00000E+0000   2.00000E+0000   3.00000E+0000
Exp6  30 Q12_missed_III UNI  3.00000E+0000   4.00000E+0000   5.00000E+0000
Exp6  31 Q13%_Multiples UNI  5.00000E+0000   1.00000E+0001   1.50000E+0001
Exp6  32 Q13a_PnextMult UNI  5.00000E+0000   1.00000E+0001   1.50000E+0001
Exp6  33 Q14a_SpatUncer uni  2.50000E+0002   3.00000E+0002   4.00000E+0002
Exp6  34 Q14RadiusError uni  8.00000E+0000   1.00000E+0001   2.00000E+0001
```

ELICITATION SESSION RESULTS
EXPERT 7 [CM=89.9% - ERF=20.9% - EW=12.5%]

```
SEED QUESTIONS
Exp7   1    CF_intervals log  1.00000E+0002   2.10000E+0002   3.00000E+0002
Exp7   2 %_vent_Serapis UNI   1.00000E+0001   5.00000E+0001   7.00000E+0001
Exp7   3      %isopachs UNI   1.00000E+0001   8.00000E+0001   1.00000E+0002
Exp7   4           HCO3 log   1.00000E-0001   3.00000E+0002   5.00000E+0003
Exp7   5 Ves_10cm_isoph UNI   1.00000E+0001   4.50000E+0001   2.00000E+0002
Exp7   6      %2km_PDCs UNI   1.00000E+0001   3.00000E+0001   7.00000E+0001
Exp7   7     %trachytic UNI   1.00000E+0001   5.00000E+0001   7.00000E+0001
Exp7   8 Area_lava_flow log   5.00000E+0001   3.00000E+0002   9.00000E+0002
Exp7   9     Vol_magma UNI    3.00000E+0003   9.00000E+0003   2.00000E+0004
Exp7  10 Area_CFc_depre UNI   6.00000E+0001   1.50000E+0002   3.00000E+0002
Exp7  11 Gas_ratio_diff UNI   2.00000E+0000   2.00000E+0001   7.00000E+0001
Exp7  12 Magma_Evac_tim UNI   5.00000E+0000   2.00000E+0001   4.80000E+0001
Exp7  13     Surf._drop UNI   2.00000E+0001   1.50000E+0002   3.00000E+0002
Exp7  14  Height_change UNI   3.00000E+0000   6.00000E+0001   6.00000E+0001
Exp7  15     Crit_vesic UNI   1.00000E-0001   1.00000E+0000   1.00000E+0001
Exp7  16   Time_+1degC UNI    3.00000E-0002   1.00000E+0000   6.00000E+0000
```

```
EXPERT-WISE RANGE GRAPHS OF THE SEED QUESTIONS RESPONSES

   1(l) [---------------------*----------]
Real :::::::::::::::::::::::::#::::::::::::::::::::::::::::::::::::::::::::::::::

   2(U)    [--------------------------------*---------------]
Real :::::::::::::::::::::::::#::::::::::::::::::::::::::::::::::::::::::::::::::

   3(U)    [--------------------------------------------------------*--------------]
Real ::::#:::::::::::::::::::::::::::::::::::::::::::::::::::::::::::::::::::::::::

   4(l)    [--------------------------------------*-------------]
Real :::::::::::::::::::::::::::::::::::::::#::::::::::::::::::::::::::::::::::::::

   5(U) [-------*------------------------------------]
Real ::::::::::#:::::::::::::::::::::::::::::::::::::::::::::::::::::::::::::::::::

   6(U) [------------------*------------------------------]
Real :::::::::::::::::::::::#:::::::::::::::::::::::::::::::::::::::::::::::::::::::

   7(U) [-----------------------------------*-----------------]
Real :::::::::::::::::::::::#::::::::::::::::::::::::::::::::::::::::::::::::::::::

   8(l)          [----------------------------*------------------]
Real ::::::::::::::::::::::::::::#::::::::::::::::::::::::::::::::::::::::::::::::::

   9(U)       [--------------------*-----------------------------------]
Real ::::#:::::::::::::::::::::::::::::::::::::::::::::::::::::::::::::::::::::::::

  10(U) [-------------------*------------------------------]
Real :::::::::::::::::::::::::::::::::::::::::::::#:::::::::::::::::::::::::::::::

  11(U) [-*------]
Real :#::::::::::::::::::::::::::::::::::::::::::::::::::::::::::::::::::::::::::::

  12(U) [--------*----------------]
Real :::::::::::::::::::#::::::::::::::::::::::::::::::::::::::::::::::::::::::::::

  13(U)    [----------------------*--------------------------]
Real :::::::::::::::::::::::::::::::::::::::::::::::::::::::::::::::::::::::::::::#
```

```
 14(U)                   [--------------*-----------------]
Real    ::::::::::::::::::::::::::::::::::::::#::::::::::::::::::::::::::::::::::::::

 15(U) [-----*-------------------------------------------------------------------]
Real    ::::::::::::::#:::::::::::::::::::::::::::::::::::::::::::::::::::::::::::::

 16(U) *--]
Real    :::::::::::::::::::::::::::::::::::::::#:::::::::::::::::::::::::::::::::::::
```

```
TARGET QUESTIONS
Exp7   17 Q1Wt_5variable uni  5.00000E+0001  8.00000E+0001  9.50000E+0001
Exp7   18    Q2Wt_Uniform UNI  5.00000E+0000  2.00000E+0001  5.00000E+0001
Exp7   19 Q3Wt_vents_loc UNI  4.00000E+0001  6.00000E+0001  9.00000E+0001
Exp7   20  Q4Wt_structure UNI  2.00000E+0001  4.00000E+0001  6.00000E+0001
Exp7   21     Q5Wt_I_vent UNI  1.00000E+0001  3.30000E+0001  5.00000E+0001
Exp7   22    Q6Wt_II_vent UNI  1.00000E+0001  3.30000E+0001  5.00000E+0001
Exp7   23   Q7Wt_III_vent UNI  2.00000E+0001  3.40000E+0001  7.00000E+0001
Exp7   24     Q7a_pre-AMS UNI  2.00000E+0001  4.00000E+0001  6.00000E+0001
Exp7   25     Q7b_postAMS UNI  4.00000E+0001  6.00000E+0001  8.00000E+0001
Exp7   26    Q8Wt_faults UNI  3.00000E+0001  6.00000E+0001  8.00000E+0001
Exp7   27 Q9Wt_fractures UNI  2.00000E+0001  4.00000E+0001  6.00000E+0001
Exp7   28    Q10_missed_I UNI  5.00000E+0000  7.00000E+0000  1.00000E+0001
Exp7   29  Q11_missed_II UNI  0.00000E+0000  1.00000E+0000  2.00000E+0000
Exp7   30 Q12_missed_III UNI  1.00000E+0000  2.00000E+0000  4.00000E+0000
Exp7   31  Q13%_Multiples UNI  5.00000E+0000  1.00000E+0001  2.00000E+0001
Exp7   32 Q13a_PnextMult UNI  5.00000E+0000  1.00000E+0001  2.00000E+0001
Exp7   33 Q14a_SpatUncer uni  1.50000E+0002  5.00000E+0002  1.00000E+0003
Exp7   34 Q14RadiusError uni  5.00000E+0000  1.50000E+0001  3.00000E+0001
```

ELICITATION SESSION RESULTS
EXPERT 8 [CM=0.1% - ERF=10.5% - EW=12.5%]

```
SEED QUESTIONS
Exp8    1    CF_intervals log  1.00000E+0002  2.50000E+0002  6.00000E+0002
Exp8    2 %_vent_Serapis UNI  5.00000E+0001  6.00000E+0001  8.00000E+0001
Exp8    3      %isopachs UINT  9.90000E+0000  1.50000E+0001  2.00000E+0001
Exp8    4           HCO3 log  1.00000E-0001  1.00000E+0000  1.00000E+0001
Exp8    5 Ves_10cm_isoph UNI  1.00000E+0001  5.00000E+0001  2.00000E+0002
Exp8    6      %2km_PDCs UNI  2.00000E+0001  3.00000E+0001  5.00000E+0001
Exp8    7     %trachytic UNT  1.00000E+0001  2.00000E+0001  5.00000E+0001
Exp8    8 Area_lava_flow log  2.00000E+0002  4.00000E+0002  8.00000E+0002
Exp8    9     Vol_magma UNI  2.00000E+0002  1.00000E+0003  5.00000E+0003
Exp8   10 Area_CFc_depre UNI  6.50000E+0001  1.00000E+0002  2.00000E+0002
Exp8   11 Gas_ratio_diff UNI  4.00000E+0001  2.00000E+0002  5.00000E+0002
Exp8   12 Magma Evac tim UNI  3.00000E+0000  5.00000E+0000  1.00000E+0001
Exp8   13     Surf._drop UNI  1.50000E+0001  8.00000E+0001  2.50000E+0002
Exp8   14  Height_change UNI  2.00000E-0001  1.00000E+0000  5.00000E+0000
Exp8   15     Crit_vesic UNI  7.00000E-0001  1.00000E+0000  1.30000E+0000
Exp8   16   Time_+1degC UNI  6.00000E+0000  2.40000E+0001  1.20000E+0002
```

```
EXPERT-WISE RANGE GRAPHS OF THE SEED QUESTIONS RESPONSES

  1(1) [------------------------*---------------------------]
Real    :::::::::::::::::::::::::::#::::::::::::::::::::::::::::::::::::::::::::::::

  2(U)                            [-------*---------------]
Real    :::::::::::::::::::::::::::#:::::::::::::::::::::::::::::::::::::::::::::::::

  3(U)    [--*---]
Real    ::::#::::::::::::::::::::::::::::::::::::::::::::::::::::::::::::::::::::::::

  4(1)   [-----------*-----------]
Real    ::::::::::::::::::::::::::::::#::::::::::::::::::::::::::::::::::::::::::::::

  5(U) [--------*------------------------------------]
Real    ::::::::#:::::::::::::::::::::::::::::::::::::::::::::::::::::::::::::::::::::

  6(U)          [---------*------------------]
Real    ::::::::::::::::::#::::::::::::::::::::::::::::::::::::::::::::::::::::::::::::

  7(U) [-------*-------------------------]
Real    ::::::::::::::#::::::::::::::::::::::::::::::::::::::::::::::::::::::::::::::::

  8(1)                            [----------*-----------]
Real    :::::::::::::::::::::::::::#:::::::::::::::::::::::::::::::::::::::::::::::::
```

```
  9(U)  [--*-------------]
Real   ::::#:::::::::::::::::::::::::::::::::::::::::::::::::::::::::::::::::::::

 10(U)  [-------*------------------------]
Real   :::::::::::::::::::::::::::::::::::::::::::::::::::#:::::::::::::::::::::::

 11(U)      [--------------------*----------------------------------------------]
Real   :#:::::::::::::::::::::::::::::::::::::::::::::::::::::::::::::::::::::::::

 12(U)  [*--]
Real   :::::::::::::::::#:::::::::::::::::::::::::::::::::::::::::::::::::::::::::

 13(U)      [-----------*-----------------------------]
Real   :::::::::::::::::::::::::::::::::::::::::::::::::::::::::::::::::::::::::::#

 14(U)              *-]
Real   ::::::::::::::::::::::::::::::::::::#:::::::::::::::::::::::::::::::::::::

 15(U)      [-*-]
Real   :::::::::::::::#:::::::::::::::::::::::::::::::::::::::::::::::::::::::::::

 16(U)      [----------*-----------------------------------------------------]
Real   :::::::::::::::::::::::::::::::::::::#:::::::::::::::::::::::::::::::::::::

TARGET QUESTIONS
Exp8    17 Q1Wt_5variable uni  3.00000E+0001   7.00000E+0001   9.00000E+0001
Exp8    18   Q2Wt_Uniform UNI  1.00000E+0001   3.00000E+0001   7.00000E+0001
Exp8    19 Q3Wt_vents_loc UNI  5.00000E+0001   7.00000E+0001   9.00000E+0001
Exp8    20 Q4Wt_structure UNI  2.00000E+0001   3.00000E+0001   5.00000E+0001
Exp8    21    Q5Wt_I_vent UNI  3.00000E+0001   5.00000E+0001   8.00000E+0001
Exp8    22   Q6Wt_II_vent UNI  2.00000E+0001   3.00000E+0001   5.00000E+0001
Exp8    23  Q7Wt_III_vent UNI  1.00000E+0001   2.00000E+0001   3.00000E+0001
Exp8    24    Q7a_pre-AMS UNI  5.00000E+0000   5.00000E+0001   9.50000E+0001
Exp8    25    Q7b_postAMS UNI  5.00000E+0000   5.00000E+0001   9.50000E+0001
Exp8    26    Q8Wt_faults UNI  5.00000E+0000   5.00000E+0001   9.50000E+0001
Exp8    27 Q9Wt_fractures UNI  5.00000E+0000   5.00000E+0001   9.50000E+0001
Exp8    28   Q10_missed_I UNI  4.00000E+0000   7.00000E+0000   1.40000E+0001
Exp8    29   Q11_missed_II UNI 0.00000E+0000   1.00000E+0000   3.00000E+0000
Exp8    30 Q12_missed_III UNI  1.00000E+0000   2.00000E+0000   4.00000E+0000
Exp8    31 Q13%_Multiples UNI  5.00000E+0000   5.00000E+0001   9.50000E+0001
Exp8    32 Q13a_PnextMult UNI  5.00000E+0000   5.00000E+0001   9.50000E+0001
Exp8    33 Q14a_SpatUncer uni  1.50000E+0002   3.00000E+0002   1.00000E+0003
Exp8    34 Q14RadiusError uni  5.00000E+0000   1.00000E+0001   2.50000E+0001
```

ITEM-WISE RANGE GRAPHS OF TARGET QUESTIONS RESPONSES

```
Item no.:  17   Item name: Q1Wt_5variable
Experts
  1                                               [----------------------*----]
  2                                                   [-----*----]
  3                                         [----------*----------]
  4                                              [----------*----]
  5                                 [----------------*----------------]
  6                                 [-----*----------]
  7                           [-------------------------------*----------------]
  8   [---------------------------------------------------*--------------------]
 CM                            [==========================*===============]
 ERF                             [======================*============]
 EW            [================================================*============]
      ----------------------------------------------------------------------
      30                                                                  95
```

```
Item no.:  18   Item name: Q2Wt_Uniform
Experts
  1  [----*----------------------]
  2  [----*-----]
  3  [----------*-----]
  4  [----*-----------]
  5  [----------------*-----------------]
  6            [----------*-----]
  7  [-----------------*-----------------------------------]
  8        [--------------------*----------------------------------------------]
 CM   [=================*=============================]
 ERF  [===========*=======================]
 EW   [===========*=========================================]
      ----------------------------------------------------------------------
      5                                                                    70
```

```
Item no.:  19   Item name: Q3Wt_vents_loc
Experts
  1              [-----------------------------*----------------------------]
  2             [--------------*----------------------]
  3             [------*--------------]
  4             [------------------------------*---------------]
  5          [-------------------------*----------------------]
  6      [------*-------]
  7  [---------------------------*----------------------------------------]
  8             [----------------------------*-----------------------------]
 CM   [============================*======================================]
 ERF          [========================*=====================]
 EW       [========================*==========================]
      ----------------------------------------------------------------------
      40                                                                  90
```

```
Item no.:  20   Item name: Q4Wt_structure
Experts
  1  [----------------------------*------------------------]
  2                  [--------------*---------------]
  3                        [------*--------------]
  4       [-------------*-----------------------------]
  5           [----------------------*----------------]
  6                                    [------*-------]
  7          [------------------------*----------------------------]
  8          [--------------*---------------------------]
 CM          [=============================*=========================]
 ERF             [=====================*=====================]
 EW        [======================*========================]
      ----------------------------------------------------------------------
      10                                                                  60
```

```
Item no.:   21   Item name: Q5Wt_I_vent
Experts
  1 [--------------*-------------------]
  2               [----*---------------]
  3                       [----*----]
  4     [---------*---------]
  5     [---------------*-------------]
  6        [----*----]
  7     [----------------------*---------------]
  8                 [------------------*----------------------------]
 CM     [========================*=================]
 ERF        [============*=============]
 EW      [=================*=========================================]
      ----------------------------------------------------------------
       5                                                          80
```

```
Item no.:   22   Item name: Q6Wt_II_vent
Experts
  1 [--------------------------*-----------------------------]
  2              [-----------------------*--------------]
  3                    [------*-------]
  4          [--------------------------*--------------------]
  5          [-------------------------------------------*---------------]
  6      [-------*-------]
  7 [---------------------------*-------------------------]
  8          [---------------*---------------------------]
 CM [=================================*=============================]
 ERF     [===========================*=====================]
 EW    [=============================*==============================]
      ----------------------------------------------------------------
       10                                                         60
```

```
Item no.:   23   Item name: Q7Wt_III_vent
Experts
  1              [------------------------*-----------------------]
  2              [-----------*-----------]
  3              [-----------*----------------------]
  4              [----------------*----------------------------------]
  5       [----------------*--------------------------]
  6                            [-----*-----------]
  7     [----------------*----------------------------]
  8 [-----------*-----------]
 CM     [===================*==================================]
 ERF          [==============*=====================]
 EW   [==========================*========================]
      ----------------------------------------------------------------
       10                                                         75
```

```
Item no.:   24   Item name: Q7a_pre-AMS
Experts
  1            [-------------*-----------]
  2            [---------*-------]
  3      [---*------------]
  4            [--------*-----------]
  5       [----------------*---------------]
  6                       [---*---]
  7       [----------------*---------------]
  8 [------------------------------*------------------------------]
 CM     [=================*==============]
 ERF       [============*=============]
 EW     [======================*==========================]
      ----------------------------------------------------------------
       5                                                          95
```

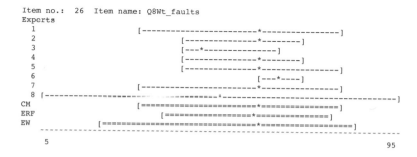

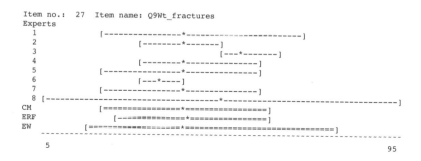

```
Item no.:  29   Item name: Q11_missed_II
Experts
   1                       [----------------------*------------------------]
   2 [-------------------------*------------------------]
   3 [-------------------------*------------------------]
   4 [-------------------------*------------------------]
   5 [-------------------------*------------------------]
   6                  [----------------------*------------------------]
   7 [-------------------------*------------------------]
   8 [-------------------------*---------------------------------------]
  CM [=========================*============================]
 ERF   [=========================*==========================]
  EW [=========================*=================================]
   -------------------------------------------------------------------
      1E-005                                                        3
```

```
Item no.:  30   Item name: Q12_missed_III
Experts
   1              [--------------*------------------------------]
   2              [--------------*--------------]
   3 [-------------*--------------]
   4              [--------------*--------------]
   5              [--------------*--------------------------]
   6                              [--------------*--------------]
   7              [--------------*--------------------------]
   8              [--------------*--------------------------]
  CM              [==============*=================================]
 ERF                [==============*=====================]
  EW       [=======================*=================================]
   -------------------------------------------------------------------
      1E-005                                                        5
```

```
Item no.:  31   Item name: Q13%_Multiples
Experts
   1   [---*----------------]
   2   [-------*------------]
   3 [*---]
   4   [---*----------------]
   5   [---*-------]
   6   [---*---]
   7   [---*-------]
   8   [----------------------------------*----------------------------]
  CM   [===*==========]
 ERF   [======*============]
  EW  [=====*========================================================]
   ---------------------------------------------------------------------
      3                                                             95
```

```
Item no.:  32   Item name: Q13a_PnextMult
Experts
   1     [---*---------------]
   2     [-------*----------]
   3 [-*---]
   4     [---*---------------]
   5     [---*-------]
   6     [---*---]
   7     [---*-------]
   8     [-----------------------------------*----------------------------]
  CM     [===*==========]
 ERF     [======*============]
  EW    [=====*===========================================]
   ---------------------------------------------------------------------
      2                                                             95
```

```
Item no.:  33  Item name: Q14a_SpatUncert
Experts
   1    [-------*----------------------------------]
   2  [----------------*--------------------]
   3  [---------*-----------------------------------------------------]
   4    [-------*----------------------]
   5  [----------------------*--------------------------]
   6        [-*-----]
   7  [------------------*------------------------]
   8  [-------*----------------------------------]
CM      [================*=========================]
ERF     [==========*==============================]
EW    [===========*====================================================]
      ----------------------------------------------------------------
     100                                                          1500
```

```
Item no.:  34  Item name: Q14RadiusError
Experts
   1 [------------*------------------------------------------]
   2 [----------------------------------*---------------------------]
   3 [------------*----------------------------]
   4 [------------*---------------------------------------]
   5 [-------------------------------*---------------]
   6         [-----*----------------------------]
   7 [------------*-------------------------------------------------]
   8 [------------*-----------------------------]
CM   [=====================================================================]
ERF   [====================*================================]
EW   [====================*==============================================]
     ----------------------------------------------------------------
      5                                                            30
```

6.3. Implemented R-codes

In the following are reported the numerical codes implemented for this study. For each script will be provided a short description (highlighted with a # symbol). Some of the codes are reported on two or more columns. The names of the scripts are reported in bold black, the input parameters are reported in blue, and the calls for other scripts are reported in red. The codes are available upon request, and should be acknowledged as Bevilacqua [2015], Doubly stochastic models for volcanic vent opening probability and pyroclastic density current hazard at Campi Flegrei caldera, PhD Thesis SNS.

6.3.1. Experts elicitation and triangular distributions

```
#Sampling of a maximum entropy distribution of given 5th, 50th, 95th percentiles and range.
max_entropy=function(incm,mid,incM,rA,rB){
x=runif(1); y=runif(1,mid,incM)
if(x>0.95){y=runif(1,incM,rB)}; if(x<0.5){y=runif(1,incm,mid); if(x<0.05){y=runif(1,rA,incm)}}
return(y)}
```

```
#Calculation of equal weights DM percentiles, given the arrays of 5th, 50th and 95th percentiles chosen by the experts.
#Monte Carlo simulation assumed n=10^5 samples.
poolDM_mixtEW=function(V_incm,V_mid,V_incM,K){        s=ceiling(runif(1)*Ne)
N=length(V_incm[1,]); Ne= length(V_incm[,1])          risp[j]=max_entropy(incm[s],mid[s],incM[s],rA,rB)}
quan05=quan50=numeric(N)                              quan05[i]=quantile(risp,0.05)
for(i in 1:N){                                        quan50[i]=quantile(risp,0.5)
incm=V_incm[,i]; mid=V_mid[,i]; incM=V_incM[,i]       quan95[i]=quantile(risp,0.95)
rA=min(incm); rB=max(incM)                            print(i); flush.console()}
R=rB-rA; rA=rA-R/10; rB=rB+R/10                        outp=matrix(0,N,3)
risp=numeric(n)                                        outp[,1]=quan05; outp[,2]=quan50; outp[,3]=quan95
for(j in 1:n){                                         return(outp)}
```

```
#Calculation of the ERF weights of the experts, given seed questions     #Defining an ordered list of DM responses.
#true values; arrays of 5th, 50th and 95th percentiles by the experts.   reOrderAns=function(DM_ans){
generate_ERF=function(true_seed,mseed,seed,Mseed){                        reDM=NULL
Ne=length(mseed[,1]); Nq=length(mseed[1,])                                reDM$A1=DM_ans[c(1,3,9,12:10,7:5),2]/100
pERF=numeric(Ne); p_single=numeric(Nq)                                    reDM$A2=DM_ans[c(2,4,8),2]/100
for(i in 1:Ne){for(j in 1:Nq){                                            reDM$M1=DM_ans[c(1,3,9,12:10,7:5),3]/100
p_single[j]=ERFweight(true_seed[j],mseed[i,j],seed[i,j],Mseed[i,j])}      reDM$M2=DM_ans[c(2,4,8),3]/100
pERF[i]=mean(p_single)}                                                   reDM$m1=DM_ans[c(1,3,9,12:10,7:5),1]/100
return(pERF/sum(pERF))}                                                   reDM$m2=DM_ans[c(2,4,8),1]/100; return(reDM)}
```

```
#ERF calculation, given a true answer and its elicitated percentiles.    #Quantile pooling of the answers; three percentiles.
ERFweight=function(x,a,b,c){                                              poolDM_quantile=function(V_incm,V_mid,V_incM,wERF){
S=NewRap(a,b,c); a=S[1]; c=S[2]                                           N=length(V_incm [1,]); outp=matrix(0,N,3)
A=min(max(0.95*x,a),c); B=max(min(1.05*x,c),a)                            outp[,1]=wERF%*%V_incm
p=((A-c)^2-(B-c)^2)/((c-a)*(c-b))                                         outp[,2]=wERF%*%V_mid
if(A<b){p=1-(B-c)^2/((c-a)*(c-b))-(A-a)^2/((b-a)*(c-a))}                   outp[,3]=wERF%*%V_incM
if(B<b){p=((B-a)^2-(A-a)^2)/((b-a)*(c-a))}}                               return(outp)}
return(p)}
```

```
#Random sample assuming a triangular distribution,                        #Random sample assuming a triangular distribution,
#given mode and 5th, 95th percentiles.                                     # given mode and range.
rtrian=function(a,b,c){                                                    rtrian_inner=function(a,b,c){
if(a==c){R=b}                                                              u=runif(1)
else {R=rtrian_inner(NewRap(a,b,c)[1],b,NewRap(a,b,c)[2])}                 if(u<((b-a)/(c-a))){x=sqrt(u*(b-a)*(c-a))+a}
return(R)}                                                                 else{x=c-sqrt((1-u)*(c-a)*(c-b))}; return(x)}
```

```
#Newton-Raphson approximation of triangular distribution range, given mode and 5th, 95th percentiles.
NewRap=function(a,b,c){                                                    InvJac=function(x,y,a,b,c){
x0=a-(c-a)/6; y0=c+(c-a)/6; x=c(x0,y0)                                     A=1.9*x+0.05*y-2*a+0.05*b
for(i in 1:5){                                                            B=0.05*(x-b); C=0.05*(y-b)
x=x-crossprod(InvJac(x[1],x[2],a,b,c),FunRap(x[1],x[2],a,b,c))}           D=1.9*y+0.05*x-2*c+0.05*b
return(x)}                                                                M=matrix(0,2,2)
                                                                          M[1,1]=D; M[2,2]=A
FunRap=function(x,y,a,b,c){                                               M[1,2]=-B; M[2,1]=-C
A1=(a-x)^2-0.05*(y-x)*(b-x); A2=(y-c)^2-0.05*(y-x)*(y-b)                   return(M/(A*D-B*C))}
return(c(A1,A2))}}
```

6.3.2. Vent opening probability

#Assembling a map given the array of probabilities of the zones of the caldera partition. We assumed dimensions 100x100 and Z
as matrix whose columns contain the indicator functions of the zones.

```
create_MapVents=function(Z,v){
M=matrix(0,100,100); nZ=length(Z[1,])
for(i in 1:nZ){ Ms=matrix(Z[,i],100,100); M=M+v[i]*Ms/sum(Ms)}
return(M/(sum(M))))}
```

#Convolution of matrix M with a gaussian kernel whose standard deviation is the average minimum distance
of n points. We assumed Mc as a matrix whose columns contain the indicator functions of such points.

```
spread_gauss=function(M,Mc,n){                     for(j in 1:100){
sigma=multi_sigma(Mc,n); m=matrix(0,100,100)       R=sqrt((h-i)^2+(k-j)^2)
for(h in 1:100){for(k in 1:100){                   m[i,j]=m[i,j]+M[h,k]*dnorm(R,0,sigma)}}}}}
if(M[h,k]>0){for( i in 1:100){                     return(m/sum(m))}
```

#Calculation of the distance between a point and a family of other points. We assume M as the indicator
function of the first point, and CC the sum of the indicator functions of all the points.

```
find_sigma=function(M,Cnt){                        if(Cnt[i,j]>0){
s=which.max(M); k=ceiling(s/100)                   R=(h-i)^2+(k-j)^2
h=s-(k-1)*100; Cnt=Cnt-sign(M); d=10000            if(d>R){d=R}}}
for(i in 1:100){for(j in 1:100){                   return(sqrt(d))}
```

#Calculation of the distance between a point and the nth closest of a family of other points. We assume M as the
indicator function of the first point, and Cnt the sum of the indicator functions of all the points.

```
find_sigma_v2=function(M,n,Cnt){                   X=Cnt[i,j]; if(X>0){
s=which.max(M); k=ceiling(s/100)                   R=(h-i)^2+(k-j)^2; flag=flag+1
h=s-(k-1)*100; Cnt=Cnt-sign(M)                     d[flag]=R; while((X-1)>0){
d=numeric(sum(Cnt)-1);flag=0                        flag=flag+1; d[flag]=R; X=X-1}}}}
for(i in 1:100){for(j in 1:100){                   d=sort(d); d=d[n]; return(sqrt(d))}
```

#Calculation of the average minimum distance of n points.

```
multi_sigma=function(Mc,n){
d=0; for(i in 1:n){cE=matrix(Mc[,i],100,100); d=d+find_sigma(cE)}
return(d/n)}
```

#Calculation of the probabilities on the nodes of the logic tree, given a list of DM responses.

```
sampleWeightsTree=function(DM_ans){                Nv=P[4:6]*100+c(29,8,33)
A1=DM_ans$A1; A2=DM_ans$A2                          P[4:6]=c(29,8,33)/Nv[1:3]
M1=DM_ans$M1; m1=DM_ans$m1                          for(j in 7:9){P[j]=rtrian(m1[j],A1[j],M1[j])}
M2=DM_ans$M2; m2=DM_ans$m2                          if(P[j]<0){P[j]=0}; if(P[j]>1){P[j]=1}}
P=A1; Q=A2                                          P[7.9]=P[7:9]*Nv[1:3]
for(j in 1:3){P[j]=rtrian(m1[j],A1[j],M1[j])}       if(((1-P[7])*(1-P[8])*(1-P[9]))==1){
if(P[j]<0){P[j]=0}; if(P[j]>1){P[j]=1}              P[7:9]=1/3}; S=P[7]+P[8]+P[9]
Q[j]=rtrian(m2[j],A2[j],M2[j])                      P[7:9]=P[7:9]/S; p=numeric(12)
if(Q[j]<0){Q[j]=0}; if(Q[j]>1){Q[j]=1}              p[1]=P[1]; p[2]=1-P[1]; p[3]=P[1]*P[2]
if((1-P[j])*(1-Q[j])==1){                           p[4]=P[1]*(1-P[2]); p[5]=P[1]*P[2]*P[7]*P[4]
P[j]=0.5; Q[j]=0.5}}                                p[6]=P[1]*P[2]*P[8]*P[5]; p[7]=P[1]*P[2]*P[9]*P[6]
P[1:3]=P[1:3]/(P[1:3]+Q[1:3])                       p[8]=P[1]*(1-P[2])*P[3]; p[9]=P[1]*(1-P[2])*(1-P[3])
for(j in 4:6){P[j]=rtrian(m1[j],A1[j],M1[j])}       p[10]=P[1]*P[2]*P[7]*(1-P[4]); p[11]=P[1]*P[2]*P[8]*(1-P[5])
if(P[j]<0){P[j]=0}}                                 p[12]=P[1]*P[2]*P[9]*(1-P[6]); return(p)}
```

#Calculation of uncertainty percentiles of the linear weights;
#Monte Carlo assuming $n=5x10^4$ samples.

```
stat_weighs=function(DM_ans,n){                    #Calculation of the linear weights, given a list of DM responses.
vect_p=matrix(0,7,n)                               sampleWeights=function(DM_ans){
for(i in 1:n){vect_p[,i]=sampleWeights(DM_ans)     if(length(DM_ans==1){return(DM_ans)}
print(i); flush.console()}; quantW=matrix(0,7,4)   E=sampleWeightsTree(DM_ans); p=numeric(7)
return(quant_pesi=seek_quantile(vect_p,7))}        p[1:5]=E[c(5:9,2)]; p[7]=E[10]+E[11]+E[12]; return(p)}
```

```
#Monte Carlo for plotting the probability distribution of each of the weights. N=10^6.
plot_weights=function(DM_ans,N){
P=numeric(7); P=weightMode(DM_ans); vect_p=matrix(0,N,7)
for(i in 1:N){vect_p[i,]= sampleWeights (DM_ans); print(i); flush.console()}
Fvent3=vect_p[,1]*100; Fvent2=vect_p[,2]*100; Fvent1=vect_p[,3]*100
DFract=vect_p[,4]*100; DFaults=vect_p[,5]*100
Homog=vect_p[,6]*100; Homog_lostvents=vect_p[,7]*100
plot(density(Fvent2,from=0),col='red', lwd=2,xlim=c(0,80), ylim=c(0,0.3))
points(P[2]*100,0,col='red',pch=19)
lines(density(Fvent3,from=0),col='darkgreen', lwd=2)
points(P[1]*100,0,col='darkgreen',pch=19)
lines(density(Fvent1,from=0),col='blue', lwd=2)
points(P[3]*100,0,col='blue',pch=19)
lines(density(DFract,from=0),col='darkviolet', lwd=2)
points(P[4]*100,0,col='darkviolet',pch=19)
lines(density(DFaults,from=0),col='darkorange3', lwd=2)
points(P[5]*100,0,col='darkorange3',pch=19)
lines(density(Homog_lostvents,from=0),col='grey55', lwd=2)
points(P[7]*100,0,col='grey55',pch=19)
lines(density(Homog,from=0), lwd=2)
points((P[6])*100,0,pch=19)}
```

```
#Sampling of a vent opening map given a list of DM responses and the maps to combine linearly. maskA2 is a  uniform map
on the whole NYT  caldera, maskB2 only inland. (v1, v2, v3) are matrices containing probability maps based on past vents
distributions.
single_sample=function(DM_ans,v1,v2,v3){
p=sampleWeights(DM_ans); return(p[3]*v1+p[2]*v2+p[1]*v3+p[4]*vfractures+p[5]*vfaults+p[6]*maskA2+p[7]*maskB2)}
```

```
#Monte Carlo simulation of average maps of vent opening and their uncertainty percentiles. Assuming dimension 100x100 and
n=5x10^4 samples.
MonteCarlo_VO=function(DM_ans,v1,v2,v3,n){
V_average=matrix(0,100,100); V_tot=matrix(0,10000,n)
for(i in 1:n){V_average=single_sample(DM_ans,v1,v2,v3,vfractures,vfaults,maskA2,maskB2)
V_tot[,i]=matrix(V_average,10000,1); print(i); flush.console()}
return(seek_quantile(V_tot,10000))}
```

```
#Calculation of mean and 5^th, 50^th and 95^th uncertainty percentiles from a sample population of N dimensional arrays.
seek_quantile=function(V_tot,N){
i05Quantile= i95Quantile=i50Quantile=imean=numeric(N)
for(j in 1:N){i05Quantile[j]=quantile(V_tot[j,],0.05)
i95Quantile[j]=quantile(V_tot [j,],0.95)
i50Quantile[j]=quantile(V_tot [j,],0.50)
imean[j]=mean(V_tot [j,])}
outp=matrix(0,N,4); outp[,1]=i05Quantile
outp[,2]=i95Quantile; outp[,3]=i50Quantile
outp[,4]=imean
return(outp)}
```

```
#Integration of the probability M on the zone of the partition.
calculate_probZone=function(M){
nZ=length(Z[1,]); vect1=numeric(nZ)
for(i in 1:nZ){vect1[i]=sum(M*matrix(Z[,i],100,100))}
return(vect1)}
```

```
#Calculation of the mean and 5^th, 95^th percentiles of the distance of two independent samples on-land. maskHD is the indicator
function  of the land.
distance_double_vents=function(DM_ans,v1,v2,v3,n){
R=numeric(n)
for(i in 1:n){M=maskHD*single_sample(DM_ans,v1,v2,v3)
M=M/sum(M); u=runif(1); S=0
for(h1 in 1:100){for(k1 in 1:100){
S=S+M[h1,k1]; if(S>u){break}}
u=runif(1); S=0
for(h2 in 1:100){for(k2 in 1:100){
S=S+M[h2,k2]; if(S>u){break}}
R[i]=sqrt((h1-h2)^2+(k1-k2)^2)*0.250
print(i); flush.console()}
R1=quantile(R,0.05); R2=mean(R); R3=quantile(R,0.95)
print(c(R1,R2,R3)); flush.console()}
```

```
#Calculation of mean, 5^th, 50^th, 95^th percentiles of the integrated probabilities on the zones of the partition.
MonteCarlo_Table=function(DM_ans,v1,v2,v3,n){
nZ=length(Z[1,]); vW=matrix(0,nZ,n)
for(i in 1:n){M=single_sample(DM_ans,v1,v2,v3)
vW[,i]=calculate_probZone(M)
print(i); flush.console()}
return(seek_quantile(vW,nZ))}
```

```
#Calculation of mean, 5th, 50th, 95th percentiles of the integrated vent opening probabilities on CF Eastern/Western sectors, or
on-land. The div parameter imposes a subdivision for the outer caldera integrated probability.
MonteCarlo_Sectors = function(N,DM_ans,v1,v2,v3){
vW=matrix(0,5,N)
for(i in 1:N){M= single_sample(DM_ans,v1,v2,v3)
vW[1,i]=sum(M*maskHD)*100; Zeast=matrix(0,100,100)
for(i in 6:13){Zeast=Zeast+matrix(Z[,i],100,100)}
Zeast=Zeast+ matrix(Z[,17],100,100)+ matrix(Z[,14]+Z[16,],100,100)*div
vW[2,i]=sum(M*Zeast)*100; print(i); flush.console()}
return(seek_quantile(vW,2))}
```

```
#Calculation of uncertainty quantifications (relative errors between mean and percentiles), inside CI caldera. Dimensions 100x100.
indicative_unc=function(outp){
mask=maskCI
a=quantile(vectorize_caldera(matrix(1-outp[1,]/outp[4,],100,100),mask),0.95)
b=mean(vectorize_caldera(matrix(1-outp[1,]/outp[4,],100,100),mask))
c=quantile(vectorize_caldera(matrix(1-outp[1,]/outp[4,],100,100),mask),0.05)
print(-c(a,b,c)*100); flush.console()
a=quantile(vectorize_caldera(matrix(outp[2,]/outp[4,]-1,100,100),mask),0.95)
b=mean(vectorize_caldera(matrix(outp[2,]/outp[4,]-1,100,100),mask))
c=quantile(vectorize_caldera(matrix(outp[2,]/outp[4,]-1,100,100),mask),0.05)
print(c(a,b,c)*100); flush.console()}
```

```
#Calculation of uncertainty quantifications (relative errors between mean and percentiles), non-zero values inside CI caldera.
Dimensions 50x50.
indicative_unc_flux=function(outp){
mask=(1-sign(change_size(maskCI)))*abs(sign(matrix(outp[,4],50,50)))
a=quantile(vectorize_caldera(matrix(1-outp[,1]/outp[,4],50,50),mask),0.95)
b=mean(vectorize_caldera(matrix(1-outp[,1]/outp[,4],50,50),mask))
c=quantile(vectorize_caldera(matrix(1-outp[,1]/outp[,4],50,50),mask),0.05)
print(-c(a,b,c)*100); flush.console()
a=quantile(vectorize_caldera(matrix(outp[,2]/outp[,4]-1,50,50),mask),0.95)
b=mean(vectorize_caldera(matrix(outp[,2]/outp[,4]-1,50,50),mask))
c=quantile(vectorize_caldera(matrix(outp[,2]/outp[,4]-1,50,50),mask),0.05)
print(c(a,b,c)*100); flush.console()}
```

```
#Compilation of the array of the elements of M that belong to a given set whose indicator function is M2.
vectorize_caldera=function(M,M2){                    for(i in 1:N){for(j in 1:N){
N=100; if(length(M)==2500){N=50}                     if(mask[i,j]==1){vect[k]=M[i,j]; k=k+1}}}
vect=numeric(sum(M2)); k=1                            return(vect)}
```

```
#Calculation of mean, 5th, 50th, 95th uncertainty percentiles of area size where probability density exceeds assigned thresholds.
MonteCarlo_Contour=function(N){                       A[2,i]=sum(sign(sign(M-1)+1))
A=matrix(0,4,N); A0=sum(mascheraCI)/16               A[3,i]=sum(sign(sign(M-2)+1))
for(i in 1:N){                                       print(i); flush.console()}
M=single_sample(DM_ans,v1,v2,v3)*1600                A=A/16; AQ= outp=seek_quantile(A,3)
A[1,i]=sum(sign(sign(M-0.5)+1))                      return(round(AQ/A0*100,1))}
```

6.3.3. PDC areal size probability

#Calculation of the maximum likelihood lognormal parameters fitted on a sample w.r.t. epistemic uncertainty sources affecting past PDC dataset. Flag parameter choses the dataset between 5ka and 15ka, Cooke_flux contains underestimation uncertainty estimate.
```
samplePDC=function(flag){
E=Cooke_flux; E=rtrian(E[1],E[2],E[3])
if(E<0){E=0}; if(flag==1){
A=aug_flux(area_ENL5ka,E); ML0=fitdistr(A,'lognormal')
return(fitdistr(c(A,sample_small_PDC(7,ML0$estimate[1],ML0$estimate[2],0.3)),'lognormal'))}
if(flag==2){A=aug_flux(area_ENL15ka,E)
ML0=fitdistr(A,'lognormal')
f1=sample_small_PDC (12,ML0$estimate[1],ML0$estimate[2],0.3)
f2=sample_small_PDC (9,ML0$estimate[1],ML0$estimate[2],0.75)
return(fitdistr(c(A,f1,f2),'lognormal'))}}
```

#Random sampling of N lost deposits areal sizes, with a lognormal dist. of mean m and sd s, conditional on being lesser of qth quantile.
```
sample_small_PDC =function(N,m,s,q){
X=rlnorm(N,m,s); for(i in 1:N){while(X[i]>qlnorm(q,m,s)){X[i]=rlnorm(1,m,s)}}
return(X)}
```

#Conversion of radial underestimation E to areal size underestimation, for an area A.
```
aug_flux=function(A,E){E=E/1000; return(A+2*sqrt(pi*A)*E+E^2*pi)}
```

#Monte Carlo simulation for plotting the pdf and the survival function of PDC areal size, with uncertainty estimates and histograms of past events dataset; number of samples $n=10^5$.
```
plot_lognorm=function(flag,n){                       if(flag==1){
x=seq(1,2000)/5                                      plot(x,Xmean,'l',ylim=c(0,0.06), xlim=c(0,400),lwd=2)
X=matrix(0,2000,n); Y=X                              lines(x,Xmax,col='red',lwd=2); lines(x,Xmin,col='red',lwd=2)
for(i in 1:n){ML=samplePDC(flag)                     hist(area_ENL5ka,breaks=seq(0,40)*10,add=TRUE,freq=FALSE)
X[,i]=dlnorm(x, ML$estimate[1], ML$estimate[2])      dev.new(); plot(x,Ymean,'l',ylim=c(0,1), xlim=c(0,400),lwd=2)
Y[,i]=1-plnorm(x, ML$estimate[1], ML$estimate[2])    lines(x,Ymax,col='red',lwd=2); lines(x,Ymin,col='red',lwd=2)}
print(i); flush.console()}                           if(flag==2){
Xmax=Xmin=Xmean= Ymax=Ymin=Ymean=numeric(2000)       plot(x,Xmean,'l',ylim=c(0,0.06), xlim=c(0,400),lwd=2)
for(i in 1:2000){                                     lines(x,Xmax,col='blue',lwd=2); lines(x,Xmin,col='blue',lwd=2)
Xmax[i]=quantile(X[i,],0.95); Xmin[i]=quantile(X[i,],0.05)  hist(area_ENL15ka,breaks=seq(0,40)*10,add=TRUE,freq=FALSE)
Xmean[i]=mean(X[i,]); Ymax[i]=quantile(Y[i,],0.95)    dev.new(); plot(x,Ymean,'l',ylim=c(0,1), xlim=c(0,400),lwd=2)
Ymin[i]=quantile(Y[i,],0.05); Ymean[i]=mean(Y[i,])}}  lines(x,Ymax,col='blue',lwd=2); lines(x,Ymin,col='blue',lwd=2)}}
```

#Plotting of bars of past events dataset areaENL, for adding information to the survival function plot.
```
plot_BarSurvival=function(areaENL){                  if(ascis[i]>=area[j]){j=j+1; flag=flag-l}
area=c(sort(areaENL),Inf)                            ordinat[i]=flag}
L=length(area)-1; l=1/L; j=1                          lines(ascis,ordinat,xlim=c(0,400))
ascis=seq(1,10000)*0.04                               lines(rep(0,1000),seq(1,1000)/1000)
ordinat=ascis; flag=1                                 lines(seq(1,1000)*0.4,rep(0,1000))
for(i in 1:10000){                                    for(j in 1:L){lines(seq(1,1000)*aree[j]/1000,rep(1-j*l,1000))}}}
```

#Calculation of L1 distance between empirical and maximum likelihood cumulative functions; flag parameters choses between weibull and lognormal classes.
```
calculate_dist=function(Data,flag){                  Int=numeric(L+1)
plot.ecdf(Data)                                      Int[1]=integrate(F,-Inf,Data.ord[1])$value
if(flag==0){F=function(x){pweibull(x,a,s)}           F.end=function(x){1-F(x)}
a=fitdistr(Data,'weibull')$estimate[1]               Int[L+1]=integrate(F.end, Data.ord[L], Inf)$value
s=fitdistr(Data,'weibull')$estimate[2]}              for(k in 1:(L-1)){F.k=function(x){abs(F(x)-(k-0.5)/L)}
if(flag==1){F=function(x){plnorm(x,m,s)}             Int[k+1]=integrate(F.k, Data.ord[k], Data.ord[k+1])$value}
m=fitdistr(Data,'lognormal')$estimate[1]             I=sum(Int); Range=max(Data)-min(Data)
s=fitdistr(Data,'lognormal')$estimate[2])            E1=100*I/Range; print(E1); flush.console()
L=length(Data); Data.ord=sort(Data)                  return(I)}
```

#Calculates p-value of a test based on L1 distance of cumulative functions.
```
test_dist=function(Data,flag){                       for(n in 1:N){Data.rand=Ran(L)
if(flag==0){Ran=function(x){rweibull(x,a,s)}}        I.rand[n]=calculate_dist(Data.rand,flag)}
if(flag==1){Ran=function(x){rlnorm(x,m,s)}}          I.rand.ord=sort(I.rand)
N=1000; I=calculate_dist(Data,flag)                  p.value=1-which.min(abs(I.rand.ord-I))/N
I.rand=numeric(N)                                    p.value}
```

6.3.4. Box Model inverse algorithm and Monte Carlo simulation

```
#Monte Carlo simulation for PDC invasion hazard. N samples for epistemic uncertainty, M samples for physical variability of areal
sizes; flag  parameter choses the 5ka or the 15ka dataset; dem500 is a 50x50 digital elevation matrix of the CF zone.
MonteCarlo_PDC=function(N,M,v1,v2,v3,flag){X=Y=matrix(0,2500,N); Ysom=numeric(2500)
for(k in 1:N){Y[,k]=matrix(change_size(maskHD*single_sample(DM_ans,v1,v2,v3)),2500,1)
Y[,k]=Y[,k]/sum(Y[,k]); m.est=matrix(0,50,50); ML=samplePDC(flag)
for(c in 1:M){for(i in 1:50){for(j in 1:50){if(Y[i+50*(j-1),k]>0){p=Y[i+50*(j-1),k]
m.est=m.est+BM_spread(i,j,dem500,p,ML$estimate[1],ML$estimate[2])}}}}; print('S'); flush.console()}
m.est=m.est/M; print(k); flush.console(); v.est=matrix(m.est,2500,1); X[,k]=v.est}; return(seek_quantile(X,2500))}

#Scale reduction of matrix M from 100x100 matrix to 50x50 matrix.
change_size=function(M){m=matrix(0,50,50);
        for( i in 1:50){for( j in 1:50){m[i,j]=(M[2*i,2*j]+M[2*i-1,2*j]+M[2*i,2*j-1]+M[2*i-1,2*j-1])}}; return(m)}
#Calculation of a single PDC invasion through an algorithm based on Box Model. (x,y) is the vent location, dem is the digital elevation
map, p is a value multiplied to the cells invaded, mNew and SNew are parameters of the lognormal dist. of areal sizes; alpha is the
energy line initial  angle.
BM_spread=function(x,y,dem,p,mNew,sNew){              Flag=0; if(A1>area.int){Flag=1}; lx=lx1*sqrt(rapp1)
dem=dem/50000; A=rlnorm(1,mNew,sNew)                  rappv=find_second(x,y,dem,p,area.int,lx,Flag)
while(plnorm(A,mNew,sNew)>Q){A=rlnorm(1,mNew,sNew)}   lx=rappv[1]; A2=rappv[2]; rapp2=rappv[3]; lx2=lx
A=4*A; H=sqrt(A); area.int=min(ceiling(A),2304)       if(abs(1-rapp2)<0.005){
DEM=dem/tan(alpha); lx=find_cone(x,y,DEM,area.int,H)  return(matrix(rappv[4:length(rappv)],50,50))}
rappv=find_first(x,y,dem,p,area.int,lx)               if(Flag){rappv=find_root(x,y,dem,p,area.int,lx2,lx1,A2,A1)}
A1=rappv[1]; rapp1=rappv[2]; lx1=lx; if(abs(1-rapp1)<0.005){  if(!Flag){rappv=find_root(x,y,dem,p,area.int,lx1,lx2,A1,A2)}
return(matrix(rappv[3:length(rappv)],50,50))}         matrix(rappv,50,50)}
#Box model direct approach, area.int is a target areal size, lx is a tentative maximum run-out  from an energy line approximation.
find_first=function(x,y,DEM,p,area.int,lx){m.est=single_spread(x,y,DEM,lx); eps=1/10^10
m.est2=(sign(m.est-eps)+1)*(1/2); m.est=m.est2*p; A1=sum(m.est2); rapp=area.int/A1; return(c(A1,rapp,m.est))}

#Box model direct approach; lx is an adjusted tentative maximum run-out from the previous approximation; flag choses if we want to
find an approximation of the target areal size from above or from below.
find_second=function(x,y,DEM,p,area.int,lx,Flag){      Flag2=0; track=track+1
track=0; Flag2=1; while(Flag2){m.est=single_spread(x,y,DEM,lx)  if(abs(1-rapp)<0.005){return(c(lx,A2,rapp,m.est))}
eps=1/10^10; m.est2=(sign(m.est-eps)+1)*(1/2)          if(Flag){if(A2>area.int){Flag2=1} lx=lx*R^(min(3,track))}}
m.est=m.est2*p; A2=sum(m.est2)                         if(!Flag){if(A2<area.int){Flag2=1; lx=lx*R^(min(3,track))}}}
rapp=area.int/A2; if(!track){R=sqrt(rapp)}             c(lx,A2,rapp,m.est)}}

#Box model inverse approach, based on the secant method; (a,b) are maximum run-out starting points, one approximation from above
and  the other  from below; (Aa,Ab) are the associated areal sizes.
find_root=function(x,y,DEM,p,area.int,a,b,Aa,Ab){      m.est=m.est2*p; A3=sum(m.est2); rapp3=area.int/A3
step=1;while(step>0.05){lx=a+(b-a)*(area.int-Aa)/(Ab-Aa)  if(abs(1-rapp3)<0.005){return(m.est)}
m.est=single_spread(x,y,DEM,lx)                        if(A3>area.int){Ab=A3; step=abs(b-lx); b=lx}
eps=1/10^10;m.est2=(sign(m.est-eps)+1)*(1/2)           if(A3<area.int){Aa=A3; step=abs(a-lx); a=lx}; return(m.est)}}

#Energy line inverse approach, DEM is a rescaled digital elevation map, H is an arbitrary column height (rescaled by tan(alpha)).
find_cone=function(x,y,DEM,area.int,H){m.est=matrix(Inf,50,50); m.est[x,y]=DEM[x,y]; c=1; c2=sqrt(2)
m.est[x+1,y]=DEM[x+1,y]+c; m.est[x+1,y+1]=DEM[x+1,y+1]+c2; m.est[x,y+1]=DEM[x,y+1]+c
m.est[x-1,y]=DEM[x-1,y]+c; m.est[x-1,y-1]=DEM[x-1,y-1]+c2; m.est[x,y-1]=DEM[x,y-1]+c
m.est[x+1,y-1]=DEM[x+1,y-1]+c2; m.est[x-1,y+1]=DEM[x-1,y+1]+c2; L1=min(2*max(1,floor(H)),50); for(i in 2:L1){c=i; c2=sqrt(2)*i
if(x+i<50){m.est[x+i,y]=max(DEM[x+i,y]+c,m.est[x+i-1,y])}; if(x-i>1){m.est[x-i,y]=max(DEM[x-i,y]+c,m.est[x-i+1,y])}
if(y+i<50){m.est[x,y+i]=max(DEM[x,y+i]+c,m.est[x,y+i-1])}; if(y-i>1){m.est[x,y-i]=max(DEM[x,y-i+1]+c,m.est[x,y-i+1])}
for(j in 1:(i-1)){if(i^2+j^2<4*ceiling(H^2)){ c=sqrt(i^2+j^2)
if(x+i<50){if(y+j<50){m.est[x+i,y+j]=max(DEM[x+i,y+j]+c,min(m.est[x+i-1,y+j],m.est[x+i-1,y+j-1]))}}
if(x+j<50){if(y+i<50){m.est[x+j,y+i]=max(DEM[x+j,y+i]+c,min(m.est[x+j-1,y+i-1],m.est[x+j-1,y+i-1]))}}
if(x+i<50){if(y-j>1){m.est[x+i,y-j]=max(DEM[x+i,y-j]+c,min(m.est[x+i-1,y-j],m.est[x+i-1,y-j+1]))}}
if(x+j<50){if(y-i>1){m.est[x+j,y-i]=max(DEM[x+j,y-i]+c,min(m.est[x+j,y-i+1],m.est[x+j-1,y-i+1]))}}
if(x-i>1){if(y+j<50){m.est[x-i,y+j]=max(DEM[x-i,y+j]+c,min(m.est[x-i+1,y+j],m.est[x-i+1,y+j-1]))}}
if(x-j>1){if(y+i<50){m.est[x-j,y+i]=max(DEM[x-j,y+i]+c,min(m.est[x-j,y+i-1],m.est[x-j+1,y+i-1]))}}
if(x-i>1){if(y-j>1){m.est[x-i,y-j]=max(DEM[x-i,y-j]+c,min(m.est[x-i+1,y-j],m.est[x-i+1,y-j+1]))}}
if(x-j>1){if(y-i>1){m.est[x-j,y-i]=max(DEM[x-j,y-i]+c,min(m.est[x-j,y-i+1],m.est[x-j+1,y-i+1]))}}}}
if(i<2*ceiling(H/sqrt(2))){if(x+i<50){if(y+i<50){m.est[x+i,y+i]=max(DEM[x+i,y+i]+c2,m.est[x+i-1,y+i-1])}}
if(y-i>1){if(x+i<50){m.est[x+i,y-i]=max(DEM[x+i,y-i]+c2,m.est[x+i-1,y-i+1])}};
if(x-i>1){if(y-i>1){m.est[x-i,y-i]=max(DEM[x-i,y-i]+c2,m.est[x-i+1,y-i+1])}}
if(y+i<50){if(x-i>1){m.est[x-i,y+i]=max(DEM[x-i,y+i]+c2,m.est[x-i+1,y+i-1])}}}}
eps=1/10^7; vect=sort.int(m.est, area.int); true.H=vect[area.int]+eps; m.est=(sign(true.H-m.est)+1)*(1/2); return(vect[area.int])}}
```

6.3.5. Dirichlet uncertainty and simultaneous eruptions

```
# Monte Carlo simulation for PDC invasion hazard, Dirichlet uncertainty model based on Selva et al. [2012].
MonteCarlo_PDCdirichlet=function(N,M){
X=Y=matrix(0,2500,N); Ysom=numeric(2500)
for(k in 1:N){
Y[,k]=matrix(change_size (maskHD *trasfGrid(matrix(sampleDirichlet(),28,25))),2500,1)
Y[,k]=Y[,k]/sum(Y[,k]); m.est=matrix(0,50,50); ML=samplePDC(flag)
for(c in 1:M){for(i in 1:50){for(j in 1:50){
if(Y[i+50*(j-1),k]>0){p=Y[i+50*(j-1),k]
m.est=m.est+BM_spread(i,j,dem500,p,ML$estimate[1],ML$estimate[2])
print(c(i,j,c,k)); flush.console()}}}}
m.est=m.est/M; v.est=matrix(m.est,2500,1); X[,k]=v.est}
return(seek_quantile(X,2500))}
```

```
#Transformation of M from Selva et al. [2012] coordinates to Bevilacqua et al. [2015] coordinates.
trasfGrid=function(M){
M1=matrix(0,50,50)
for(i in 12:39){for(j in 7:31){
M1[i,j]=M[i-11,j-6]}}; M2=matrix(0,100,100)
for(i in 1:100){for(j in 1:99){
M2[i,j]=M1[ceiling((i)/2),ceiling((j+1)/2)]}}
return(M2)}
```

```
#Dirichlet distribution sample, assuming 700 dimensions and
coefficients in A.post.
generaDirichlet=function(){
T.inc=numeric(700)
for(j in 1:700){T.inc[j]=rgamma(1,A.post[j])}
T.inc=T.inc/sum(T.inc)
return(T.inc)}
```

```
#Vent opening sample assuming only past vents dataset.
single_sampleOnlyVents=function(DM_ans,vents1,vents2,vents3){
p=sampleWeights(DM_ans)
S=p[1]+p[2]+p[3]; p=p/S
V_average =p[3]*vents1+p[2]*vents2+p[1]*vents3
return(V_average)}
```

```
# Monte Carlo simulation for PDC invasion hazard, double eruptions; Cooke_double is the elicited probability (with uncertainty
percentiles) of a double eruption conditioned to an eruption.
MonteCarlo_PDCdouble=function(N,M,v1,v2,v3,flag){
X=Y=matrix(0,2500,N); Ysom=numeric(2500)
for(k in 1:N){Y[,k]=matrix(change_size(maskHD*single_sample(DM_ans,v1,v2,v3)),2500,1)
Y[,k]=Y[,k]/sum(Y[,k]); m.est=matrix(0,50,50); ML=samplePDC(flag)
E=rtrian(Cooke_double[1], Cooke_double[2], Cooke_double[3])/100; if(E<0){E=0}
for(c in 1:M){for(i in 1:50){for(j in 1:50){
if(Y[i+50*(j-1),k]>0){p=Y[i+50*(j-1),k]
m.est=m.est+BM_spread (i,j,dem500,p,ML$estimate[1],ML$estimate[2])}}}}
print('S'); flush.console()}
m.est=m.est/M; m.est2=matrix(0,50,50)
for(c in 1:M){for(i in 1:50){for(j in 1:50){
if(Y[i+50*(j-1),k]>0){p=Y[i+50*(j-1),k]
add.mat=BM_spread(i,j,dem500,p,ML$estimate[1],ML$estimate[2])
m.est2=m.est2+add.mat+p*((1-sign(add.mat))*m.est)}}}
print('D'); flush.console()}
m.est=m.est2/M*E+m.est*(1-E)
print(k);flush.console()
v.est=matrix(m.est,2500,1); X[,k]=v.est}
return(seek_quantile(X,2500))}}
```

6.3.6. Box model parameters statistics

```
#Plotting of the mass estimates in function of the maximum run-out (without topography). w is the settling velocity,
phi0 the starting volume fraction, F the Froude number, gp is the reduced gravity.
plot_mass=function(lmax){
w=seq(1,1000)/1000*(2-0.01)+0.01; phi0=seq(1,1000)/1000*(0.02-0.0005)+0.0005; mass=matrix(0,1000,1000)
for(i in 1:1000){for(j in 1:1000){
mass[i,j]=pi*(1.22+phi0[i]*998.78)*volume(w[j],F,gp,phi0[i],lmax); contour(mass)
C=1; lines(seq(1,1000)/1000,(8*C^3)/(11329*phi0),col='blue', lwd=2, lty=2)
C=1.8; lines(seq(1,1000)/1000,(8*C^3)/(11329*phi0),col='darkviolet', lwd=2, lty=2)
C=2; lines(seq(1,1000)/1000,(8*C^3)/(11329*phi0),col='red', lwd=2)
C=2.4; lines(seq(1,1000)/1000,(8*C^3)/(11329*phi0),col='red', lwd=2,lty=2)}

#Volume estimation for each radiant.
volume=function(w,F,gp,f,lx){return((lx*8*w^2/f/F^2/gp)^(1/3)/4)}

#Plotting of the energy estimates in function of the distance from the vent (without topography).
plot_energy=function(lmax){
x=seq(1,1000)/1000*lmax
C=1; plot(x,(C*(lmax)^(1/3)/(x/lmax*cosh(atanh((x/lmax)^2))^2))^2/(2*g),col='blue', lwd=2, 'l',ylim=c(0,1000)  )
C=1.8; lines(x,(C*(lmax)^(1/3)/(x/lmax*cosh(atanh((x/lmax)^2))^2))^2/(2*g),col='darkviolet', lwd=2, lty=2)
C=2; lines(x,(C*(lmax)^(1/3)/(x/lmax*cosh(atanh((x/lmax)^2))^2))^2/(2*g),col='red', lwd=2)
C=2.4; lines(x,(C*(lmax)^(1/3)/(x/lmax*cosh(atanh((x/lmax)^2))^2))^2/(2*g),col='red', lwd=2, lty=2)}

#Generic function for plotting matrix H.
plotta=function(H){image(-H,col=heat.colors(1000))}

#plotting of the C constant probability distribution.
distributionC=function(N){
C=numeric(N); for(i in 1:N){
w=runif(1)*(1.2_0.06)+0.06; F=runif(1)*(1.19-1)+1
f=runif(1)*(0.015-0.005)+0.005; r=runif(1)*(1000-700)+700
g=(r-1.22)/1.22*9.81; C[i]= constantC(w,F,g,f)
print(i); flush.console()}
print(c(quantile(C,0.05),mean(C),quantile(C,0.95))); flush.console()
print(quantile(C,0.5)); flush.console()
plot.density(density(C), lwd=2)
lines(rep(1,10000),seq(1,10000)/10000*1.5, lty=2, col='blue', lwd=2)
lines(rep(1.8,10000),seq(1,10000)/10000*1.5, lty=2, col='darkviolet', lwd=2)
lines(rep(2,10000),seq(1,10000)/10000*1.5, col='red', lwd=2)
lines(rep(2.4,10000),seq(1,10000)/10000*1.5, lty=2, col='red', lwd=2)}

#C constant calculation.
constantC=function(w,F,g,f){return((w*F^2*g*f)^(1/3)/2)}
```

6.3.7. Box model direct propagation algorithm

```
#Simple circular direct propagation.
circle_spread=function(x,y,p){
r=rlnorm(1,m,s); m.est=matrix(0,50,50)
while(plnorm(r,m,s)>Q){r=rlnorm(1,m,s)}
r=sqrt(r/pi); R=r/0.5; i=0
for(i in 0:floor(R)){j=0; while ((i^2+j^2)<(R^2)){
        if(x+i<51){if(y+j<51){m.est[x+i,y+j]=m.est[x+i,y+j]+p}}
        if(y-j>0){if(j>0){m.est[x+i,y-j]=m.est[x+i,y-j]+p}}}
        if(x-i>0){if(i>0){if(y+j<51){m.est[x-i,y+j]=m.est[x-i,y+j]+p}
        if(y-j>0){if(j>0){m.est[x-i,y-j]=m.est[x-i,y-j]+p}}}}
        j=j+1}}; return(m.est)}
```

```
#Box model direct propagation algorithm, contiguous cells approach (cells crossed by the line from the actual position to the vent).
#Size parameter is the length of each cell aside (Method 1).
single_spread=function(x,y,DEM,lx){
S=25000/size; m.est=matrix(-Inf,S,S); m.est[x,y]=Inf; c=decay(1,lx); c2=decay(sqrt(2),lx)
m.est[x+1,y]=c-DEM[x+1,y]; m.est[x+1,y+1]=c2-DEM[x+1,y+1]; m.est[x,y+1]=c-DEM[x,y+1]
m.est[x-1,y]=c-DEM[x-1,y]; m.est[x-1,y-1]=c2-DEM[x-1,y-1]; m.est[x,y-1]=c-DEM[x,y-1]
m.est[x+1,y-1]=c2-DEM[x+1,y-1]; m.est[x-1,y+1]=c2-DEM[x-1,y+1]; L2=min(max(1,floor(lx)),S)
for(i in 2:L2){c=decay(i,lx)
if(x+i<S){m.est[x+i,y]=min(c-DEM[x+i,y],m.est[x+i-1,y])}
if(x-i>1){m.est[x-i,y]=min(c-DEM[x-i,y],m.est[x-i+1,y])}
if(y+i<S){m.est[x,y+i]=min(c-DEM[x,y+i-1],m.est[x,y+i-1])}
if(y-i>1){m.est[x,y-i]=min(c-DEM[x,y-i+1],m.est[x,y-i+1])}
for(j in 1:(i-1)){if(i^2+j^2<ceiling(lx^2)){c=decay(sqrt(i^2+j^2),lx)
if((x+i<S)*(y+j<S)){m.est[x+i,y+j]=min(c-DEM[x+i,y+j],max(m.est[x+i-1,y+j],m.est[x+i-1,y+j-1]))}}
if((x+j<S)*(y+i<S)){m.est[x+j,y+i]=min(c-DEM[x+j,y+i],max(m.est[x+j,y+i-1],m.est[x+j-1,y+i-1]))}}
if((x+i<S)*(y-j>1)){m.est[x+i,y-j]=min(c-DEM[x+i,y-j],max(m.est[x+i-1,y-j],m.est[x+i-1,y-j+1]))}}
if((x+j<S)*(y-i>1)){m.est[x+j,y-i]=min(c-DEM[x+j,y-i],max(m.est[x+j,y-i+1],m.est[x+j-1,y-i+1]))}}
if((x-i>1)*(y+j<S)){m.est[x-i,y+j]=min(c-DEM[x-i,y+j],max(m.est[x-i+1,y+j],m.est[x-i+1,y+j-1]))}}
if((x-j>1)*(y+i<S)){m.est[x-j,y+i]=min(c-DEM[x-j,y+i],max(m.est[x-j+1,y+i],m.est[x-j+1,y+i-1]))}}
if((x-i>1)*(y-j>1)){m.est[x-i,y-j]=min(c-DEM[x-i,y-j],max(m.est[x-i+1,y-j],m.est[x-i+1,y-j+1]))}}}}
if((x-j>1)*(y-i>1)){m.est[x-j,y-i]=min(c-DEM[x-j,y-i],max(m.est[x-j,y-i+1],m.est[x-j+1,y-i+1]))}}}}}
c2=decay(sqrt(2)*i,lx); if(i<ceiling(lx/sqrt(2))){
if((x+i<S)*(y+i<S)){m.est[x+i,y+i]=min(c2-DEM[x+i,y+i],m.est[x+i-1,y+i-1])}}
if((y-i>1)*(x+i<S)){m.est[x+i,y-i]=min(c2-DEM[x+i,y-i],m.est[x+i-1,y-i+1])}}
if((x-i>1)*(y-i>1)){m.est[x-i,y-i]=min(c2-DEM[x-i,y-i],m.est[x-i+1,y-i+1])}}
if((y+i<S)*(x-i>1)){m.est[x-i,y+i]=min(c2-DEM[x-i,y+i],m.est[x-i+1,y+i-1])}}}}}
return(m.est)}
```

```
#Alternative contiguous cells assumption (any cell having its centre closer to the vent than the actual position).
single_spread_v2=function(x,y,DEM,lx){
S=25000/size; m.est=matrix(-Inf,S,S); m.est[x,y]=Inf; c=decay(1,lx); c2=decay(sqrt(2),lx)
m.est[x+1,y]=c-DEM[x+1,y]; m.est[x+1,y+1]=c2-DEM[x+1,y+1]; m.est[x,y+1]=c-DEM[x,y+1]
m.est[x-1,y]=c-DEM[x-1,y]; m.est[x-1,y-1]=c2-DEM[x-1,y-1]; m.est[x,y-1]=c-DEM[x,y-1]
m.est[x+1,y-1]=c2-DEM[x+1,y-1]; m.est[x-1,y+1]=c2-DEM[x-1,y+1]; L2=min(max(1,floor(lx)),S)
for(i in 2:L2){c=decay(i,lx)
if(x+i<S){m.est[x+i,y]=min(c-DEM[x+i,y],max(m.est[x+i-1,y],m.est[x+i-1,y-1],m.est[x+i-1,y+1]))}
if(x-i>1){m.est[x-i,y]=min(c-DEM[x-i,y],max(m.est[x-i+1,y],m.est[x-i+1,y-1],m.est[x-i+1,y+1]))}
if(y+i<S){m.est[x,y+i]=min(c-DEM[x,y+i-1],max(m.est[x,y+i-1],m.est[x-1,y+i-1],m.est[x+1,y+i-1]))}
if(y-i>1){m.est[x,y-i]=min(c-DEM[x,y-i+1],max(m.est[x,y-i+1],m.est[x-1,y-i+1],m.est[x+1,y-i+1]))}
for(j in 1:(i-1)){if(i^2+j^2<ceiling(lx^2)){c=decay(sqrt(i^2+j^2),lx)
if((x+i<S)*(y+j<S)){m.est[x+i,y+j]=min(c-DEM[x+i,y+j],max(m.est[x+i-1,y+j],m.est[x+i-1,y+j-1],m.est[x+i,y+j-1],m.est[x+i-1,y+j+1]))}}
if((x+j<S)*(y+i<S)){m.est[x+j,y+i]=min(c-DEM[x+j,y+i],max(m.est[x+j,y+i-1],m.est[x+j-1,y+i-1],m.est[x+j-1,y+i],m.est[x+j+1,y+i-1]))}}
if((x+i<S)*(y-j>1)){m.est[x+i,y-j]=min(c-DEM[x+i,y-j],max(m.est[x+i-1,y-j],m.est[x+i-1,y-j+1],m.est[x+i,y-j+1],m.est[x+i-1,y-j-1]))}}
if((x+j<S)*(y-i>1)){m.est[x+j,y-i]=min(c-DEM[x+j,y-i],max(m.est[x+j,y-i+1],m.est[x+j-1,y-i+1],m.est[x+j-1,y-i],m.est[x+j+1,y-i+1]))}}
if((x-i>1)*(y+j<S)){m.est[x-i,y+j]=min(c-DEM[x-i,y+j],max(m.est[x-i+1,y+j],m.est[x-i+1,y+j-1],m.est[x-i,y+j-1],m.est[x-i+1,y+j+1]))}}
if((x-j>1)*(y+i<S)){m.est[x-j,y+i]=min(c-DEM[x-j,y+i],max(m.est[x-j,y+i-1],m.est[x-j+1,y+i-1],m.est[x-j+1,y+i],m.est[x-j-1,y+i-1]))}}
if((x-i>1)*(y-j>1)){m.est[x-i,y-j]=min(c-DEM[x-i,y-j],max(m.est[x-i+1,y-j],m.est[x-i+1,y-j+1],m.est[x-i,y-j+1],m.est[x-i+1,y-j-1]))}}}}
if((x-j>1)*(y-i>1)){m.est[x-j,y-i]=min(c-DEM[x-j,y-i],max(m.est[x-j,y-i+1],m.est[x-j+1,y-i+1],m.est[x-j+1,y-i],m.est[x-j-1,y-i+1]))}}}}}
c2=decay(sqrt(2)*i,lx); if(i<ceiling(lx/sqrt(2))){
if((x+i<S)*(y+i<S)){m.est[x+i,y+i]=min(c2-DEM[x+i,y+i],max(m.est[x+i-1,y+i-1],m.est[x+i,y+i-1],m.est[x+i-1,y+i]))}}
if((y-i>1)*(x+i<S)){m.est[x+i,y-i]=min(c2-DEM[x+i,y-i],max(m.est[x+i-1,y-i+1],m.est[x+i,y-i+1],m.est[x+i-1,y-i]))}}
if((x-i>1)*(y-i>1)){m.est[x-i,y-i]=min(c2-DEM[x-i,y-i],max(m.est[x-i+1,y-i+1],m.est[x-i,y-i+1],m.est[x-i+1,y-i]))}}
if((y+i<S)*(x-i>1)){m.est[x-i,y+i]=min(c2-DEM[x-i,y+i],max(m.est[x-i+1,y+i-1],m.est[x-i,y+i-1],m.est[x-i+1,y+i]))}}}}}
return(m.est)}
```

```
#Decay energy function calculation.
decay=function(x,lx){decad=0; if(x<lx){decad=(C*(size*lx)^(1/3)/(x/lx*cosh(atanh((x/lx)^2))^2))^2/(2*g)/size}; return(decad)}
```

```
#Direct PDC propagation, shading alternative approach (Method 2).
single_spread_shade=function(x,y,DEM,lx){
S=25000/size; L2=min(max(1,floor(lx)),S)+1; m.est=matrix(-Inf,S,S)
A1=max(x-L2,1); A2=min(x+L2,S); B1=max(y-L2,1); D2=min(y+L2,S)
m.est_red=m.est[A1:A2,B1:B2]; DEMr=DEM[A1:A2,B1:B2]
xred=x-A1; yred=y-B1; M1=A2-A1+1; M2=B2-B1+1
for(i in 1:M1){for(j in 1:M2){
m.est_red[i,j]=decay(sqrt((i-xred)^2+(j-yred)^2),lx)-DEMr[i,j]}
print(i); flush.console()}; m.est_redFIX=m.est_red

for(i in 1:M1){for(j in 1:M2){
if(m.est_redFIX[i,j]<0){
m.est_red=shade(xred,yred,i,j,M1,M2,m.est_red)}}
print(i); flush.console()}
m.est[A1:A2,B1:B2]=m.est_red
m.est[x,y]=Inf; eps=1/10^12
m.est=(sign(m.est-eps)+1)*(1/2)
return(m.est)}
```

```
#Shading alg. on the submatrix M1xM2 of M, centered on (x,y).
shade=function(x,y,i,j,M1,M2,M){
A=i-x; B=j-y; if((A*B)==0){
if(A>0){M[i:M1,]=M[i:M1,]*produce_shade2(A,B,M1-i+1,M2,j)}
if(B>0){M[,j:M2]=M[,j:M2]*produce_shade2(A,B,M1,M2-j+1,i)}
if(A<0){M[1:i,]=M[1:i,]*produce_shade2(A,B,i,M2,j)}
if(B<0){M[,1:j]=M[,1:j]*produce_shade2(A,B,M1,j,i)}}
if(A*B!=0){if(A>0){if(B>0){M[i:M1,j:M2]=M[i:M1,j:M2]*produce_shade(A,B,M1-i+1,M2-j+1)}
if(B<0){M[i:M1,1:j]=M[i:M1,1:j]*produce_shade(A,B,M1-i+1,j)}}
if(A<0){if(B>0){M[1:i,j:M2]=M[1:i,j:M2]*produce_shade(A,B,i,M2-j+1)}
if(B<0){M[1:i,1:j]=M[1:i,1:j]*produce_shade(A,B,i,j)}}}; return(M)}
```

```
#Simple conversion of digital elevation map data.
modDEM=function(){
DEMtrue=DEM10m
for(i in 1:2500){for(j in 1:2500){
if(DEM10m[i,j]==-9999){DEMtrue[i,j]=0}}}
return(DEMtrue)}
```

```
#angular shade (case 1)
produce_shade=function(x,y,H,K){
flag=2*sign(x)+sign(y)
if(flag==3){R1=(x+0.5)/(y-0.5); R2=(x-0.5)/(y+0.5)
M=matrix(1,H,K); for(i in 1:H){for(j in 1:K){R=(i+x-1)/(j+y-1)
if((R1>R)*(R2<R)){M[i,j]=0}}}}
if(flag==-1){R1=(-x+0.5)/(y-0.5);R2=(-x-0.5)/(y+0.5)
M=matrix(1,H,K); for(i in 1:H){for(j in 1:K){R=(i-x-1)/(j+y-1)
if((R1>R)*(R2<R)){M[H-i+1,j]=0}}}}
if(flag==1){R1=(x+0.5)/(-y-0.5); R2=(x-0.5)/(-y+0.5)
M=matrix(1,H,K); for(i in 1:H){for(j in 1:K){R=(i+x-1)/(j-y-1)
if((R1>R)*(R2<R)){M[i,K-j+1]=0}}}}
if(flag==-3){R1=(-x+0.5)/(-y-0.5), R2=(-x-0.5)/(-y+0.5)
M=matrix(1,H,K); for(i in 1:H){for(j in 1:K){R=(i-x-1)/(j-y-1)
if((R1>R)*(R2<R)){M[H-i+1,K-j+1]=0}}}}; return(M)}
```

```
#central shade (case 2)
produce_shade2=function(x,y,H,K,I){
flag=2*sign(x)+sign(y)
if(flag==2){R1=0.5/(x-0.5); R2=(-0.5)/(x-0.5)
M=matrix(1,H,K); for(i in 1:H){for(j in 1:K){R=(I-j)/(x+i-1)
if((R1>R)*(R2<R)){M[i,j]=0}}}}
if(flag==-2){R1=0.5/(-x-0.5); R2=(-0.5)/(-x-0.5)
M=matrix(1,H,K); for(i in 1:H){for(j in 1:K){R=(I-j)/(-x+i-1)
if((R1>R)*(R2<R)){M[H-i+1,j]=0}}}}
if(flag==1){R1=0.5/(y-0.5); R2=(-0.5)/(y-0.5)
M=matrix(1,H,K); for(i in 1:K){for(j in 1:H){R=(I-j)/(y+i-1)
if((R1>R)*(R2<R)){M[j,i]=0}}}}
if(flag==-1){R1=0.5/(-y-0.5); R2=(-0.5)/(-y-0.5)
M=matrix(1,H,K); for(i in 1:K){for(j in 1:H){R=(I-j)/(-y+i-1)
if((R1>R)*(R2<R)){M[j,K+i-1]=0}}}}; return(M)}
```

```
#Direct PDC propagation on radial sectors (Method 3), size=50m.
single_spread_slices=function(x,y,lx){
M=matrix(0,500,500); L2=min(max(1,floor(lx)),500)+1
A1=max(x-L2,1); A2=min(x+L2,500)
B1=max(y-L2,1); B2=min(y+L2,500)
M_red=M[A1:A2,B1:B2]
M1=A2-A1+1; M2=B2-B1+1
xred=x-A1; yred=y-B1
C1=500-xred+1; C2=500+M1-xred
D1=500-yred+1; D2=500+M2-yred
M_sr=M_slices[C1:C2,D1:D2]; M_tr=M_circles[C1:C2,D1:D2]
for(i in 0:359){j=i/360*2*pi; d=0; X=x*5-2; Y=y*5-2; K=1
while(K>0){X=X+2.5*cos(j); Y=Y+2.5*sin(j); d=d+0.5
if((X<1)+(Y<1)+(X>2500)+(Y>2500)){break}
K=decay(d,lx)-DEMtrue[X,Y]/5000}
M.est=(1-abs(sign(M_sr-i)))*(sign(sign(d-1-M_tr)+1))
print(i); flush.console(); M_red=M_red+M.est}
M[A1:A2,B1:B2]=M_red
return(M)}
```

```
#Definition of matrix containing Euclidean distances from vent.
CreateCircles=function(){
M=matrix(0,999,999)
for(i in 1:999){for(j in 1:999){
x=i-500; y=j-500
M[i,j]=sqrt(x^2+y^2)}}
return(round(M))}
```

```
# Definition of matrix containing angles w.r.t. a clockwise rotation
#around the vent (in position [500,500]).
CreateSlices=function(){
M=matrix(0,999,999)
for(i in 1:999){for(j in 1:999){
x=i-500; y=j-500; k=atan(y/x)
if(x>0){if(y>=0){M[i,j]=k}
if(y<0){M[i,j]=2*pi+k}}
if(x<0){if(y>=0){M[i,j]=pi+k}
if(y<0){M[i,j]=pi+k}}}
M=round(M/(2*pi)*360); M[500,500]=0
for(i in 1:999){for(j in 1:999){
if(M[i,j]==360){M[i,j]=0}}}
return(M)}
```

6.3.8. Temporal uncertainty modelling

6.3.8   Temporal uncertainty modelling

#random sampling of epistemic uncertainties affecting past events record (order-time-volume-localization); T data frame
contains the information about ordered events, u about the most uncertain events.

```
sampleTemp=function(T,u){
L=length(T$Id)
Times=numeric(L)
for(i in 1:L){
if(!length(T)){return(NULL)}
#stratigraphic record
A=T$Swap[i]
if(A>0){
if(A<1){
a=runif(1)
if(a<A){
T$Id[i:(i+1)]=T$Id[(i+1):i]
T$Loc[i:(i+1)]=T$Loc[(i+1):i]
T$V[i:(i+1)]=T$V[(i+1):i]
T$Tmin[i:(i+1)]=T$Tmin[(i+1):i]
T$Tmax[i:(i+1)]=T$Tmax[(i+1):i]
T$EW[i:(i+1)]=T$EW[(i+1):i]}
T$Swap[i:(i+1)]=c(0,0)}}
if(A==1){T$Swap[i]=-1}
#spatial location
A=T$Loc[i]
if(floor(A)<A){
a=runif(1)*100
if(T$Id[i]=='AMS'){
Av=c(23,77)
Bv=c(8,9)}
if(T$Id[i]=='PP'){
Av=c(48,41,11)
Bv=c(8,9,10)}
if(T$Id[i]=='S4_s31'){
Av=c(41,42,17)
Bv=c(8,9,10)}
if(T$Id[i]=='S4_s32'){
Av=c(42,33,25)
Bv=c(8,9,10)}
if(T$Id[i]=='CASA'){
Av=c(7,42,51)
Bv=c(6,9,12)}
if(T$Id[i]=='PIGN1'){
Av=c(13,4,35,38,10)
Bv=c(6,7,8,9,12)}
j=1
S=0
while(a>Av[j]+S){
S=Av[j]+S
j=j+1}
T$Loc[i]=Bv[j]}
#most uncertain VDRE (unif)
A=-T$V[i]
if(A>0){
a=runif(1)
if(A==100){
B=10+a*90}
if(A==300){
B=100+a*200}
```

```
if(A==10){
B=a*10}
T$V[i]=round(B)}
#other VDRE (trian)
if(A<0){
a=rtrian(0.5,1,1.5)
B=-A*a
T$V[i]=round(B)}}

#uncertain times (trian2)
flag1=0
while(flag1==0){
flag1=1
B0=Inf
for(i in 1:L){
A=T$Tmin[i]
if(A>0){
C=T$Tmax[i]
B=rtrian2(A,(A+C)/2,C)
Times[i]=round(B)
if(B>B0){flag1=0}
B0=B}}}

#simultaneous events
for(i in 1:L){
if(T$Swap[i]==-1){
Times[i]=Times[i-1]}}

#other times (small unif)
i=1
while(i<(L+1)){
D=Times[i]
if(D==0){
C=0
j=i
while(C==0){
j=j+1
C=Times[j]
if(j>L){C=Times[i-1]-runif(1)*100*(j-i)}
}
if(i>1){
A=Times[i-1]}
if(i==1){
A=C+runif(1)*100*(j-i)}
B=C
U=sort(round(runif(j-i,B,A)),TRUE)
Times[i:(j-1)]=U
i=j}
if(D>0){i=i+1}}

#most uncertain times (large trian2)
nId=c(Inf,as.vector(T$Id),-Inf)
nLoc=c(Inf,T$Loc,-Inf)
nEW=c(Inf,T$EW,-Inf)
nT=c(Inf,Times,-Inf)
```

```
nV=c(Inf,T$V,-Inf)
l=length(nId)-1
if(length(u)!=0){
l=length(u$Id)
uTimes=numeric(l)
flag=0
B0=Inf
while(flag==0){
flag=1
for(i in 1:l){
A=u$Tmin[i]
C=u$Tmax[i]
B=rtrian2(A,(A+C)/2,C)
uTimes[i]=round(B)
if(B>B0){flag=0}
B0=B}}
for(i in 1:l){
A=-u$V[i]
if(A>0){
a=runif(1)
if(A==100){
B=10+a*90}
if(A==300){
B=100+a*200}
if(A==10){
B=a*10}
u$V[i]=round(B)}
if(A<0){
a=rtrian(0.5,1,1.5)
B=-A*a
u$V[i]=round(B)}}
for(i in 1:l){
j=1
S=0
A=nT[j]
while(uTimes[i]<A){
A=nT[j+1]
j=j+1}
l=length(nId)
nId=c(nId[1:(j-1)],u$Id[i],nId[j:l])
nLoc=c(nLoc[1:(j-1)],u$Loc[i],nLoc[j:l])
nEW=c(nEW[1:(j-1)],u$EW[i],nEW[j:l])
nT=c(nT[1:(j-1)],uTimes[i],nT[j:l])
nV=c(nV[1:(j-1)],u$V[i],nV[j:l])}}}

T=NULL
T$Id=nId[2:l]
T$Loc=nLoc[2:l]
T$EW=nEW[2:l]
T$T=nT[2:l]
T$V=nV[2:l]
T= as.data.frame(T)
T$Id=as.vector(T$Id)
return(T)}
```

```
#Monte Carlo simulation for computing the duration of epochs and inter-epochs with uncertainty; N samples.
DuratEpo=function(TI,TII,TIII,N,uI,uIII){            for(j in 1:N){for(i in 1:6){
l1=length(TI$Id); l2=length(TII$Id); l3=length(TIII$Id)    Ebounds[i]=c(rtrian2(vDm[i],(vDM[i]+vDm[i])/2,vDM[i]))}
vDm=c(TIII$Tmin[c(l3,1)],TII$Tmin[c(l2,2)],TI$Tmin[c(l1-1,3)])    A=rtrian2(uI$Tmin[1],(uI$Tmin[1]+uI$Tmax[1])/2,uI$Tmax[1])
vDm[4]=vDm[4]+runif(1)*100                            B=rtrian2(uI$Tmin[2],(uI$Tmin[2]+uI$Tmax[2])/2,uI$Tmax[2])
vDm[5]=vDm[5]-runif(1)*100                            C=rtrian2(uIII$Tmin[1],(uIII$Tmin[1]+uIII$Tmax[1])/2,uIII$Tmax[1])
vDm[6]=vDm[6]+2*runif(1)*100                          Ebounds[1]=min(Ebounds[1],C)
vDM=c(TIII$Tmax[c(l3,1)],TII$Tmax[c(l2,2)],TI$Tmax[c(l1-1,3)])    Ebounds[5]=min(Ebounds[5],A,B)
vDM[4]=vDM[4]+runif(1)*100                            Ebounds[6]=max(Ebounds[6],A,B)
vDM[5]=vDM[5]-runif(1)*100                            Eduration[,j]=Ebounds[2:6]-Ebounds[1:5]
vDM[6]=vDM[6]-2*runif(1)*100                          if(j/100==floor(j/100)){print(j); flush.console()}}
Ebounds=numeric(6)                                    return(round(seek_quantile(Eduration,5)))}
Eduration=matrix(0,5,N)
```

```
#Simple Monte Carlo sampling for computing time between Nisida and Monte Nuovo eruptions.
Inter3=function(N){
Times=numeric(N)
for(i in 1:N){Times[i]=rtrian2(3213,(3213+4188)/2,4188)-477
if((i/100)==floor(i/100)){print(i); flush.console()}}
A=c(quantile(Times,0.05),mean(Times),quantile(Times,0.95))
print(A);flush.console()}
```

```
#Random sampling with a triangular distribution, Newton-Raphson approximation of range, given mode and 2.5th, 97.5th percentiles.
rtrian2=function(a,b,c){if(a==c){R=b}            NewRap2=function(a,b,c){
else {R=rtrian_inner(NewRap2(a,b,c)[1],b,NewRap2(a,b,c)[2])}    x0=a-(c-a)/6; y0=c+(c-a)/6
return(R)}                                        x=c(x0,y0); for(i in 1:5){
                                                  x=x-crossprod(InvJac2(x[1],x[2],a,b,c),FunRap2(x[1],x[2],a,b,c))}
InvJac2=function(x,y,a,b,c){                      return(x)}
A=1.95*x+0.025*y-2*a+0.025*b
B=0.025*(x-b); C=0.025*(y-b)                       FunRap2=function(x,y,a,b,c){
D=1.95*y+0.025*x-2*c+0.025*b                       A1=(a-x)^2-0.025*(y-x)*(b-x)
M=matrix(0,2,2); M[1,]=c(D,-B)                     A2=(y-c)^2-0.025*(y-x)*(y-b)
M[2,]=c(-C,A); return(M/(A*D-B*C))}                return(c(A1,A2))}
```

```
#random new eruption times sampling              #random past eruption times sampling,
sample_timesHawk=function(T,mu,I0,DL){            #optional T time window start and DL duration.
DataSample=matrix(0,DL,16)                        sample_pastTimes=function(mT,u,T=0,DL=0){
zV=spaFreHawk(NULL,NULL,NULL)                      TM=sampleTemp(mT,u)
for(i in 1:16){if(zV[i]>0){                        N=length(TM$Id)
I0_loc=I0*zV[i]                                    if(!DL){DL=TM$T[1]-TM$T[N]+5}
k=3/T; h=mu*k; I_clust=0                            DataSample=matrix(0,DL,13)
if(i==3){I_clust=h}                                if(!T){T=TM$T[1]+1}
tau=0; while(tau<DL){                              TM$T=-TM$T+T
I_ini=I_clust; test=1; dt=0                         if(TM$T[1]<0){return(0)}
while(test<(I_clust+I0_loc)){                       for(i in 1:13){
dt=dt+rexp(1,I_ini+I0_loc)                           vM=LocDatHawk(TM,i,0)
I_clust=I_ini*exp(-k*dt)                             if(length(vM)>0){
test=runif(1)*(I_ini+I0_loc)}                        for(k in 1:length(vM)){
tau=tau+dt; I_clust=I_clust+h                         if(vM[k]<DL){DataSample[vM[k],i]=1}}}}
if(tau<DL){                                         Y=seq(1,DL)
DataSample[floor(tau),i]=1}}}                        plot(rep(1,DL),Y,'l',xlim=c(1,13),lwd=2)
Y=seq(1,DL)                                          for(i in 2:13){
plot(rep(1,DL),Y,'l',xlim=c(1,13),lwd=2)             lines(rep(i,DL),Y,lwd=2)}
for(i in 2:13){lines(rep(i,DL),Y,lwd=2)}             for(j in 1:13){for(i in 1:DL){
for(j in 1:13){for(i in 1:DL){                       if(DataSample[i,j]>0){
if(DataSample[i,j]>0){                               if(N==33){points(j,i,col='blue',pch=3,lwd=2)}
points(j,i,col='red',pch=3,lwd=2)}}}                 if(N==28){points(j,i,col='darkgreen',pch=3,lwd=2)}}}}
return(sum(DataSample))}                             sum(DataSample)}
```

6.3.9. Volume sampling and plotting

6.3.9   Volume sampling and plotting

#Monte Carlo and plotting of volume estimates with uncertainty; N samples, localized in n1:n2 zones, X1 and X2 parameter set bounds (if 0 default), flag1=0 if create new plot and 1,2 if we add to the current. If flag=1 plots events, if flag=0 volumes estimates.

```
volPlot=function(N,T,n1,n2,X1,X2,u,flag1,flag){
Out=CrQuant(T,u,N,n1,n2,flag)
S=Out$S; x=Out$x; L=Out$L
if(X1==0){X1=max(x)}; if(X2==0){X2=min(x)}
l=n2-n1; if(flag1==0){if(L==28){
plot(-x,S[,2],'l',col='darkgreen',lty=2,xlim=c(-X1,-X2))
lines(-x,S[,1],col='darkgreen',lty=2)
lines(-x,S[,4],lwd=2,col='darkgreen'); lines(-x,S[,4])
if(l==15){lines(-x,S[,3],col='darkgreen')}}
if(L==8){; plot(-x,S[,2],'l',col='red',lty=2,xlim=c(-X1,-X2))
lines(-x,S[,1],col='red',lty=2)
lines(-x,S[,4],lwd=2,col='red'); lines(-x,S[,4])
if(l==15){lines(-x,S[,3],col='red')}}
if(L==33){plot(-x,S[,2],'l',col='blue',lty=2,xlim=c(-X1,-X2))
lines(-x,S[,1],col='blue',lty=2)
lines(-x,S[,4],lwd=2,col='blue'); lines(-x,S[,4])
if(l==15){lines(-x,S[,3],col='blue')}}}
if(flag1==1){if(L==28){
```
```
lines(-x,S[,2],col='forestgreen',lty=2)
lines(-x,S[,1],col='forestgreen',lty=2)
lines(-x,S[,4],lwd=2,col='forestgreen'); lines(-x,S[,4])}
if(L==8){lines(-x,S[,2],lty=2,col='indianred')
lines(-x,S[,1],col='indianred',lty=2)
lines(-x,S[,4],lwd=2,col='indianred'); lines(-x,S[,4])}
if(L==33){lines(-x,S[,2],col='dodgerblue3',lty=2)
lines(-x,S[,1],col='dodgerblue3',lty=2)
lines(-x,S[,4],lwd=2,col='dodgerblue3'); lines(-x,S[,4])}}
if(flag1==2){if(L==28){lines(-x,S[,2],col='green3',lty=2)
lines(-x,S[,1],col='green3',lty=2)
lines(-x,S[,4],lwd=2,col='green3'); lines(-x,S[,4])}
if(L==8){lines(-x,S[,2],lty=2,col='chocolate')
lines(-x,S[,1],col='chocolate',lty=2)
lines(-x,S[,4],lwd=2,col='chocolate'); lines(-x,S[,4])}
if(L==33){lines(-x,S[,2],col='deepskyblue',lty=2)
lines(-x,S[,1],col='deepskyblue',lty=2)
lines(-x,S[,4],lwd=2,col='deepskyblue'); lines(-x,S[,4])}}}}
```

#Monte Carlo and plotting of total volume estimates or event number with uncertainty; N samples, flag1 plots events, flag=0 volumes estimates. # n1:n2 for localizing in zones, flag1=0 if create new plot and 1 if we add to the current.

```
volPlot2=function(N,T1,T2,T3,u1,u3,flag,n1=-Inf,n2=Inf, flag1=0){
Out=CrQuant(T1,u1,N,-Inf,Inf,flag); S1=Out$S; x1=Out$x; Out=CrQuant(T2,NULL,N,-Inf,Inf,flag); S2=Out$S; x2=Out$x
Out=CrQuant(T3,u3,N,-Inf,Inf,flag); S3=Out$S; x3=Out$x; F1=min(x1); F2=min(x2); F3=min(x3); I2=max(x2); I3=max(x3)
X1=seq(0,1000)/1000*(I2-F1)+F1; X2=seq(0,1000)/1000*(I3-F2)+F2; X3=seq(0,1000)/1000*(477-F3)+F3;
                    X3bis=seq(0,1000)/1000*(0-477)+477
v=c(0,0,0); if(n1<4){if(n2>2){v=c(15,30,45)/1000)}; if(flag){v=rep(1,3)}; A1='blue'; A2='red'; A3='darkgreen'
if(!flag1){plot(-X1,rep(max(S1[,4]),1001),ylim=c(0,max(S1[,2])+max(S2[,2])+max(S3[,2])+v[3]),xlim=c(-max(x1),0),'l',lty=2)}
if(flag1){lines(-X1,rep(max(S1[,4]),1001),lty=2);  A1='dodgerblue3'; A2='indianred'; A3='forestgreen'}
lines(-X1,rep(max(S1[,2]),1001),lty=2); lines(-X1,rep(max(S1[,1]),1001),lty=2); lines(-X2,rep(max(S2[,4]),1001),lty=2)
lines(-X2,rep(max(S2[,1])+max(S1[,1]),1001),lty=2); lines(-X2,rep(max(S2[,2])+max(S1[,2]),1001),lty=2)
lines(-X3,rep(max(S3[,1])+max(S2[,1])+max(S1[,1]),1001),lty=2); lines(-X3,rep(max(S3[,2])+max(S2[,2])+max(S1[,2]),1001),lty=2)
lines(-X3,rep(max(S3[,4])+max(S2[,4])+max(S1[,4]),1001),lty=2); lines(-
lines(-X3bis,rep(max(S3[,1])+max(S2[,1])+max(S1[,1])+v[1],1001),lty=2);
                    lines(-X3bis,rep(max(S3[,2])+max(S2[,2])+max(S1[,2])+v[3],1001),lty=2)
lines(-x1,S1[,1],col=A1,lty=2); lines(-x1,S1[,2],col=A1,lty=2); lines(-x1,S1[,4],lwd=2, col=A1); lines(-x1,S1[,4]); lines(-x1,S1[,3], col=A1)
lines(-x2,S2[,1]+max(S1[,1]),col=A2, lty=2); lines(-x2,S2[,2]+max(S1[,2]),col=A2, lty=2)
lines(-x2,S2[,4]+max(S1[,4]),lwd=2,col=A2); lines(-x2,S2[,4]+max(S1[,4])); lines(-x2,S2[,3]+max(S1[,3]),col=A2)
lines(-x3,S3[,1]+max(S1[,1])+max(S2[,1]),col=A3, lty=2); lines(-x3,S3[,2]+max(S1[,2])+max(S2[,2]),col=A3, lty=2)
lines(-x3,S3[,4]+max(S1[,4])+max(S2[,4]),lwd=2,col=A3); lines(-x3,S3[,4]+max(S1[,4])+max(S2[,4]))
lines(-x3,S3[,3]+max(S1[,3])+max(S2[,3]),col=A3); v1=seq(0,100)/100*v[1]; v2=seq(0,100)/100*v[3]; v3=seq(0,100)/100*v[2]
if(n1<4){if(n2>2){lines(rep(-477,101),v2+max(S3[,2])+max(S1[,2])+max(S2[,2]),lty=2,col='darkviolet')
lines(rep(-477,101),v1+max(S3[,1])+max(S1[,1])+max(S2[,1]),lty=2,col='darkviolet')
lines(rep(-477,101),v3+max(S3[,4])+max(S1[,4])+max(S2[,4]),lwd=2,col='darkviolet')
lines(rep(-477,101),v3+max(S3[,3])+max(S1[,3])+max(S2[,3]),col='darkviolet')
lines(rep(-477,101),v3+max(S3[,4])+max(S1[,4])+max(S2[,4]))}}}}
```

#Calculating the time domain and the volume quantiles.

```
CrQuant=function(T,u,N,n1,n2,flag1){
L=length(T$Id)+length(u$Id); Vol=matrix(0,L,N)
Time=Vol; for(i in 1:N){MT=sampleTemp(T,u)
for(j in 1:L){if(MT$Loc[j]<=n2){if(MT$Loc[j]>=n1){
Vol[j,i]=MT$V[j]}}}; Time[,i]=MT$T
if(flag1){Vol[,i]=rep(1,L)}
print(i); flush.console()}
xlim1=max(Time[1,]); xlim2=min(Time[L,])
x=-seq(0,10000)/10000*(xlim1-xlim2)+xlim1
y=matrix(0,10001,N)
```
```
for(k in 1:N){i=1; s=0
flag=1; for(j in 1:10001){y[j,k]=s
if(x[j]<Time[i,k]){if((i*flag)==L){
s=Vol[i,k]+s; y[j,k]=s; flag=0}
if(i<L){s=Vol[i,k]+s; y[j,k]=s; i=i+1}}}
print(k); flush.console()}
S=seek_quantile(y,10001)/1000
if(flag1){S=S*1000}
Out=NULL; Out$x=x; Out$S=S
Out$L=L; return(Out)}
```

6.3.10. Maximum likelihood parameters for Hawkes processes

```
#Calculating optimal (n,T) in [n1,n2] x [T1,T2], double iteration scales, if MC=0 printing more info;
#mT are data sample of epoch zV; example - zV=3 for epoch III, zV=1 for epoch I
#option EW=1 for considering only Western sector, EW=2 for Eastern sector. Option Split=1 for considering only events before AMS/PP.
#optionally given also mT2 and mT3 - for merging with mT, changing zV - example zV=12 for merging epoch I and II
MaxLikHawk=function(n1,n2,T1,T2,mT,sT,sn,zV,MC=0,f=1,EW=0,Split=0,mT2=NULL,mT3=NULL,MN=NULL){
if(Split*EW==1){return(NULL)}; if(Split*length(mT3)){return(NULL)}
if(length(mT2)){mT=MergeHawk(mT,mT2,mT3,MN)}
if(Split){for(i in 1:length(mT$Id)){if(mT$Id[i]=='PP'){Split=i}
if(mT$Id[i]=='AMS'){Split=i}}
if(Split==1){Split=0}}
zT=TotDatHawk(mT,Split); A=calcEHawk(mT,EW,Split)
N=A[1]; E=A[2]; a=A[3]; b=A[4]; l=A[5]; if(!E){return(NULL)}
zV=SpatGlobHawk(zV,mT,mT2,mT3,a,b);
return(InnLikHawk(n1,n2,T1,T2,zT,mT,sT,sn,f,E,a,b,l,N,zV,MC))
```

```
#Defining a matrix with all past localized event times before event L
TotDatHawk=function(T,L=0){
if(!length(T$Id)){return(NULL)}
data=matrix(NA,16,40)
for(i in 1:16){
S=LocDatHawk(T,i,L)
if(length(S)){
data[i,1:length(S)]=S}}
return(data)}
```

```
#Defining the random spatial frequencies
SpatGlobHawk=function(zV,mT,mT2=NULL,mT3=NULL,a=1,b=16){
if(min(zV)==1){zV=SpaFreHawk(mT,NULL,NULL)}
if(min(zV)==2){zV=SpaFreHawk(NULL,mT,NULL)}
if(min(zV)==3){zV=SpaFreHawk(NULL,NULL,mT)}
if(min(zV)==12){zV=SpaFreHawk(mT,mT2,NULL)}
if(min(zV)==123){zV=SpaFreHawk(mT,mT2,mT3)}
if(min(zV)==1234){zV=SpaFreHawk(mT,mT2,mT3)}
zV=zV/70*69; zV[3]=zV[3]+1/70}
zV=c(rep(0,a-1),zV[a:b],rep(0,16-b))
zV=zV/sum(zV); return(zV)}
```

```
#Inner algorithm for repeating two iterations at different scales
InnLikHawk=function(n1,n2,T1,T2,zT,mT,sT,sn,f,E,a,b,l,N,zV,MC){
if(n1<0){n1=0}; if(n2>N){n2=N}
dom_T=(seq(((T2-T1)/sT+1)-1)*sT+T1; if(!E){return(NULL)}
dom_n=(seq(((n2-n1)/sn+1)-1)*sn+n1
dom=(n2-n1)/sn+1; nn=(T2-T1)/sT+1; L=-Inf
for(m in 1:dom){l0=dom_n[m]/E; L2=-Inf
for(j in 1:nn){k_act=3/dom_T[j]; h_act=(N-dom_n[m])/N*k_act
L1=LikHawk_zones(zT,mT,zV,l0,k_act,h_act,E,a,b,l)
if(L1>L){L=L1; T=dom_T[j]; n=dom_n[m]; k=k_act; h=h_act}
if(!MC){if(L1>L2){L2=L1}}
if(!MC){print(c(dom_n[m],L2)); flush.console()}}
if(!f){if(!MC){print(c('n =',round(n,1)),quote=F); flush.console()
print(c('T =',round(T)), quote=F); flush.console()
print(c('mu =',round((N-n)/N,2)), quote=F); flush.console()
print(c('1/l0 =',round(1/l0,2)), quote=F); flush.console()}
return(c(n,T,(N-n)/N,l0,k,h,E))}
if(f){if(T>sT){return(InnLikHawk(n-sn,n+sn,T-sT,T+sT,zT,mT,sT/20,sn/10,0,E,a,b,l,N,zV,MC))}
if(T==sT){return(InnLikHawk(n-sn,n+sn,T-(sT*9/10),T+(sT*9/10),zT,mT,sT/20,sn/10,0,E,a,b,l,N,zV,MC))}}}
```

```
#Merging of the data in a single epoch
MergeHawk=function(mT1,mT2=NULL,mT3=NULL,MN=NULL){
mT=NULL; mT$Id=c(mT1$Id,mT2$Id,mT3$Id,MN$Id)
mT$Loc=c(mT1$Loc,mT2$Loc,mT3$Loc,MN$Loc)
mT$EW=c(mT1$EW,mT2$EW,mT3$EW,MN$EW)
mT$T=c(mT1$T,mT2$T,mT3$T,MN$T)
mT$V=c(mT1$V,mT2$V,mT3$V,MN$V)
return(as.data.frame(mT))}
```

```
#Defining a vector with localized event times in zone n before event L
LocDatHawk=function(T,n,L){
if(!L){L=length(T$Id)}
data=NULL; j=1
for(i in 1:L){
if(T$Loc[i]==n){
data[j]=T$T[i]
j=j+1}}
return(data)}
```

```
#Calculating local frequencies, optional fixed samples
SpaFreHawk=function(T1,T2,T3){
if(!length(T1)){T1=sampleTemp(TI,uI)}
if(!length(T2)){T2=sampleTemp(TII,uII)}
if(!length(T3)){T3=sampleTemp(TIII,uIII)}
S=c(T1$Loc,T2$Loc,T3$Loc)
N=numeric(16)
for(i in 1:length(S)){
N[S[i]]=N[S[i]]+1}
return(N/sum(N))}
```

```
#Calculating epoch duration restricted to eastern or western
sector calcEHawk=function(mT,EW,N=0){
if(!N){N=length(mT$Id)}
if(!length(mT)){return(c(0,0,1,16,1))}
if(!EW){return(c(N,mT$T[1]-mT$T[N],1,16,1))}
n=0; S=0
for(i in 1:N){
if(mT$EW[i]==EW){n=n+1; S=i}}
l=1; while(mT$EW[l]==EW){l=l+1}
a=1; b=5
if(EW==2){a=6;b=16}
return(c(n,mT$T[l]-mT$T[S],a,b,l))}
```

```
#Calculating log-likelihood with given parameters; zones a:b.
LikHawk_zones=function(zT,mT,zV,l0,k_act,h_act,E,a,b,l=1){
if(!E){return(0)}; L1=0
for(i in a:b){time_z=zT[i,]
time_z=time_z[!is.na(time_z)]
nz=length(time_z); dataT=-time_z+(mT$T[i])
if(nz>1){dt=dataT[2:nz]-dataT[1:(nz-1)]}
if(nz<2){dt=0}
l0_act=zV[i]*l0
L1_act=seekLikeHawk(l0_act,k_act,h_act,dataT,dt,E)
L1=L1+L1_act}
return(L1)}
```

```
#Localized log-likelihood calculation with fixed l0, k, h parameters;
#dataT are increasing times, dt are time intervals, E epoch duration
seekLikeHawk=function(l0,k,h,dataT,dt,E){
n=length(dataT)
if(n==0){return(-E*l0)}
L1=-h/k*n-E*l0+h/k*sum(exp(-k*(E-dataT)))
L2=l0; l_clust=0
if(n>1){for(i in 1:(n-1)){
l_clust=(l_clust+h)*exp(-dt[i]*k)
L2=L2*(l0+l_clust)}}
L2=log(L2); return(L1+L2)}
```

```
#Version for maximize combined likelihoods of the IxIIxIII epochs (or IxIII epochs); mu and E are referred to the third epoch;
#option Merge=1 for merging epochs I and II, and then calculating the likelihood of (I*II)xIII epochs.
MaxLikHawk2=function(n1,n2,T1,T2,mT1,mT2,mT3,sT,sn,zV=0,f=1,EW=0,Split=0,Merge=0){
if(Split*EW==1){return(NULL)}; S1=S3=0; if(Merge){mT1=MergeHawk(mT1,mT2); mT2=NULL}
if(Split){for(i in 1:length(mT1$Id)){if(mT1$Id[i]=='PP'){S1=i}}
for(i in 1:length(mT3$Id)){if(mT3$Id[i]=='AMS'){S3=i}}}
zT1=TotDatHawk(mT1,S1); zT2=TotDatHawk(mT2); zT3=TotDatHawk(mT3,S3);
A=calcEHawk(mT1,EW,S1); N1=A[1]; E1=A[2]; a=A[3]; b=A[4]; I1=A[5]
A=calcEHawk(mT2,EW); N2=A[1]; E2=A[2]; I2=A[5]; if(f){zV=SpatGlobHawk(123,mT1,mT2,mT3,a,b)}
A=calcEHawk(mT3,EW,S3); N3=A[1]; E3=A[2]; I3=A[5]; if(!E3){return(NULL)}
return(InnLikHawk2(n1,n2,T1,T2,zT1,mT1,zT2,mT2,zT3,mT3,sT,sn,f,E1,E2,E3,a,b,I1,I2,I3,N1,N2,N3,zV))}
```

```
#Inner algorithm for repeating two iterations at different scales
InnLikHawk2=function(n1,n2,T1,T2,zT1,mT1,zT2,mT2,zT3,mT3,sT,sn,f,E1,E2,E3,a,b,I1,I2,I3,N1,N2,N3,zV){
if(n1<0){n1=0}; if(n2>N3){n2=N3}; dom_T=(seq((T2-T1)/sT+1)-1)*sT+T1
dom_n=(seq((n2-n1)/sn+1)-1)*sn+n1; dom=(n2-n1)/sn+1; nn=(T2-T1)/sT+1; L=-Inf
for(m in 1:dom){l0=dom_n[m]/E3
for(j in 1:nn){k_act=3/dom_T[j]; h_act=(N3-dom_n[m])/N3*k_act
L11=LikHawk_zones(zT1,mT1,zV,l0,k_act,h_act,E1,a,b,I1)
L12=LikHawk_zones(zT2,mT2,zV,l0,k_act,h_act,E2,a,b,I2)
L13=LikHawk_zones(zT3,mT3,zV,l0,k_act,h_act,E3,a,b,I3)
if(L11+L12+L13>L){L=L11+L12+L13; T=dom_T[j]; n=dom_n[m]; k=k_act; h=h_act}}}
if(!f){return(c(n,T,(N3-n)/N3,l0,k,h,E3))}
if(f){if(T>sT){return(InnLikHawk2(n-sn,n+sn,T-sT,T+sT,zT1,mT1,zT2,mT2,zT3,mT3,sT/20,sn/10,0,E1,E2,E3,a,b,I1,I2,I3,N1,N2,N3,zV))}
if(T==sT){return(InnLikHawk2(n-sn,n+sn,T-
                 (sT*9/10),T+(sT*9/10),zT1,mT1,zT2,mT2,zT3,mT3,sT/20,sn/10,0,E1,E2,E3,a,b,I1,I2,I3,N1,N2,N3,zV))}}}
```

6.3.11. Cox-Hawkes processes Monte Carlo simulations

```
#Monte Carlo (MC) simulation, N=2500 samples; inputs described in the sequel
#output contains number n of base rate events, time T of decay, mean offspring mu, base rate value I0, parameters k and h,
epoch duration E.
MonteCarloHawk=function(N,n1,n2,T1,T2,mT,u,sT,sn,zV,EW=0,Split=0,mT2=NULL,u2=NULL,mT3=NULL,u3=NULL,MN=NULL){
A=matrix(0,7,N); for(i in 1:N){
A[,i]=MaxLikHawk(n1,n2,T1,T2,sampleTemp(mT,u),sT,sn,zV,1,1,EW,Split,sampleTemp(mT2,u2),sampleTemp(mT3,u3),
                 sampleTemp(MN,NULL))
print(i); flush.console()}; outp=seek_quantile(A,7)[,c(1,4,2)]
print(c('n =',round(outp[1,],1)),quote=F); flush.console()
print(c('T =',round(outp[2,])),quote=F); flush.console()
print(c('mu =',round(outp[3,],2)),quote=F); flush.console()
print(c('1/I0 =',round(1/outp[4,c(3,2,1)],2)),quote=F); flush.console(); return(outp)}

#MC simulation - combined likelihood of more epochs, N=2500 samples
MonteCarloHawk2=function(N,n1,n2,T1,T2,mT1,u1,mT2,u2,mT3,u3,sT,sn,EW=0,Split=0,Merge=0){
A=matrix(0,7,N); for(i in 1:N){
A[,i]=MaxLikHawk2(n1,n2,T1,T2,sampleTemp(mT1,u1),sampleTemp(mT2,u2),sampleTemp(mT3,u3),sT,sn,0,1,EW,Split,Merge)
print(i); flush.console()}
outp=seek_quantile(A,7)[,c(1,4,2)]
print(c('n =',round(outp[1,],1)),quote=F); flush.console()
print(c('T =',round(outp[2,])),quote=F); flush.console()
print(c('mu =',round(outp[3,],2)),quote=F); flush.console()
print(c('1/I0 =',round(1/outp[4,c(3,2,1)],2)),quote=F); flush.console()
return(outp)}
```

```
#MC on N=500000 samples - offspring probability
MC_offspringHawk=function(N,T,Vmu){
S=numeric(3)
for(i in 1:N){
S[1]=S[1]+erupt01hawk(T,Vmu[1])
S[2]=S[2]+erupt01hawk(T,Vmu[2])
S[3]=S[3]+erupt01hawk(T,Vmu[3])
if(i/100==floor(i/100)){
print(i); flush.console()}}
return(round(S/N,2))}
```

```
#Sampling of indicator 0-1 if the eruption has offspring
erupt01hawk=function(T,mu){
D=T/3*7; k=3/T; h=mu*k
I_clust=h; test=1; dt=0
while(test>I_clust){
dt=dt+rexp(1,h)
if(dt>D){return(0)}
I_clust=h*exp(-k*dt)
test=runif(1)*h}
return(1)}
```

```
#Sample of cluster sizes
clustDistHawk=function(T,mu){
D=T/3*7; k=3/T; h=mu*k
I_clust=h; L=1; DL=h*exp(-k*D)
while(I_clust>=DL){
I_ini=I_clust; test=1; dt=0
while(test>I_clust){dt=dt+rexp(1,I_ini)
I_clust=I_ini*exp(-k*dt)
if(I_clust<DL){return(L)}
```

```
test=runif(1)*I_ini}
L=L+1; I_clust=I_clust+h}
print('FAIL'); flush.console()}
```

```
#Clusters counting
NclustHawk=function(Vn,VP){
S=numeric(3); S[1]=Vn[1]*VP[1]
S[2]=Vn[2]*VP[2]; S[3]=Vn[3]*VP[3]
return(round(S,1))}
```

```
#Clusters size (not considering 'clusters' of
#one element)
size_clustHawk=function(mu,P){
size=1/(1-mu)
trusize=(size-1+P)/P
return(round(trusize,1))}
```

```
#MC on N samples – clusters distribution
MC_clustDistHawk=function(N,T,Vmu){
S1=S2=S3=numeric(N)
for(i in 1:N){
S1[i]=clustDistHawk(T,Vmu[1])
S2[i]=clustDistHawk(T,Vmu[2])
S3[i]=clustDistHawk(T,Vmu[3])
if(i/100==floor(i/100)){
print(i); flush.console()}}
```

```
L=max(S1); Vs=numeric(L)
for(i in 1:N){
Vs[S1[i]]=Vs[S1[i]]+1}
Vs=Vs/sum(Vs)*100
A=round(Vs[1:10],1)
L=max(S2); Vs=numeric(L)
for(i in 1:N){
Vs[S2[i]]=Vs[S2[i]]+1}
Vs=Vs/sum(Vs)*100
```

```
B=round(Vs[1:10],1)
L=max(S3); Vs=numeric(L)
for(i in 1:N){
Vs[S3[i]]=Vs[S3[i]]+1}
Vs=Vs/sum(Vs)*100
C=round(Vs[1:10],1)
D=matrix(0,10,3)
D[,1]=A; D[,2]=B; D[,3]=C
return(as.data.frame(D))}
```

```
#Nested MC on NxM samples – MN offspring
MC_MNoffspringHawk=function(N,M,VT,Vmu,t0){
Vs=numeric(M); for(j in 1:M){mu=rtrian(Vmu[1],Vmu[2],Vmu[3])
T=rtrian(VT[1],VT[2],VT[3]); S=0
for(i in 1:N){S=S+MNerupt01hawk(T,mu,t0)}
Vs[j]=S/N; if(j/10==floor(j/10)){print(j); flush.console()}}
Vs1=quantile(Vs,0.05); Vsm=mean(Vs); Vs2=quantile(Vs,0.95)
return(round(c(Vs1,Vsm,Vs2)*100,2))}
```

```
#Sampling of indicator 0-1 if Monte Nuovo (MN) will have offspring
MNerupt01hawk=function(T,mu,t0){
if(mu<=0){return(0)}; if(T<=0){return(0)}
D=T/3*10; k=3/T; h=mu*k; I_clust=h*exp(-k*t0)
I_ini=I_clust; test=1; dt=0
while(test>I_clust){dt=dt+rexp(1,h)
if(dt>D){return(0)}; I_clust=I_ini*exp(-k*dt)
test=runif(1)*h}; return(1)}
```

```
#Nested MC on NxM samples – probability of no events after MN.
MC_IVepoHawk=function(N,M,VT,Vmu,VI0){
Vs=numeric(M); for(j in 1:M){S=0
mu=rtrian(Vmu[1],Vmu[2],Vmu[3])
T=rtrian(VT[1],VT[2],VT[3]); I0=rtrian(VI0[1],VI0[2],VI0[3])
for(i in 1:N){S=S+IVerupt01hawk(T,mu,I0)}
Vs[j]=S/N; print(Vs[j]); flush.console()
if(j/10==floor(j/10)){print(j); flush.console()}
Vs1=quantile(Vs,0.05); Vsm=mean(Vs); Vs2=quantile(Vs,0.95)
return(round(c(Vs1,Vsm,Vs2)*100,2))}
```

```
#Calculating uncertainty ranges
rangesHawk=function(a,b,c,f=0){
if(f){a=1/a; b=1/b; c=1/c}
A=(b-a)/b; B=(c-b)/b
print(round(100*c(-A,B),2))
flush.console()
print(round(50*(A+B),2))
flush.console()}
```

```
#Density estimation: Max. Likelihood (ML) exponential;
#parameter A imposes the color.
plotExpT=function(Vs,A=0){
mean=fitdistr(Vs,'exponential')$estimate
X=(seq(1,10001)-1)/10000*1250
if(!A){plot(X,dexp(X,mean),'l',lwd=2,xlim=c(0,1250),ylim=c(0,0.008))}
if(A==1){lines(X,dexp(X,mean),lwd=2,lty=2,col='red')}
if(A==2){lines(X,dexp(X,mean),lwd=2,lty=2,col='darkviolet')}
if(A==3){lines(X,dexp(X,mean),lwd=2,lty=2,col='blue')}
if(A==4){lines(X,dexp(X,mean),lwd=2,lty=2,col='green3')}}
```

```
#ML exponential considering East and West separately.
doublePlot=function(Vs1,Vs2){
m1=fitdistr(Vs1,'exponential')$estimate
m2=fitdistr(Vs2,'exponential')$estimate
mean=m1+m2; X=(seq(1,10001)-1)/10000*1250
plot(X,dexp(X,mean),'l',lwd=2,xlim=c(0,1250),ylim=c(0,0.008))
lines(X,dexp(X,m1),lwd=2,lty=2,col='blue')
lines(X,dexp(X,m2),lwd=2,lty=2,col='green3')}
```

```
#Sampling of indicator 0-1 that CF had no events for D years after
#MN; if f=1 samples an hypothesized repose time after MN,
#subtracting D years to the result; if f=2 samples next eruption time.
IVerupt01hawk=function(T,mu,I0,f=0,D=477){
if(mu<=0){mu=0}; if(T<=1){T=1}; k=3/T; h=mu*k
I_clust=h; if(f==2){ I_clust=h*exp(-k*D)}
I_ini=I_clust; test=1; dt=0; while(test>(I_clust+I0)){
dt=dt+rexp(1,I_ini+I0); if(!f){if(dt>D){return(1)}}
I_clust=I_ini*exp(-k*dt); test=runif(1)*(I_ini+I0)}
if(f==1){return(dt-D)}; if (f==2){return(dt)}; return(0)}
```

```
#Calculating uncertainty ranges on matrix.
rangeMatHawk=function(M){
print('T',quote=F); flush.console()
rangesHawk(M[2,1],M[2,2],M[2,3])
print('mu',quote=F); flush.console()
rangesHawk(M[3,1],M[3,2],M[3,3])
print('1/I0',quote=F); flush.console()
rangesHawk(M[4,3],M[4,2],M[4,1],1)}
```

```
#Kernel density estimation for time distrib. reflexed in zero;
#parameter A imposes the color.
plot_kdeT=function(Vs,A=0){
X=numeric(2048); m=quantile(Vs,0.995)
pdf=density(Vs,n=4096,from=-m,to=m)
X=c(pdf$y[2049:4095]+pdf$y[2048:2],pdf$y[4096])
print(sum(X*(pdf$x[1]-pdf$x[2]))); flush.console(); if(!A){
plot(pdf$x[2049:4096],X,'l',lwd=2,xlim=c(0,1250),ylim=c(0,0.008))}
if(A==1){lines(pdf$x[2049:4096],X,lwd=2,lty=2,col='red')}
if(A==2){lines(pdf$x[2049:4096],X,lwd=2,lty=2,col='darkviolet')}
if(A==3){lines(pdf$x[2049:4096],X,lwd=2,lty=2,col='blue')}
if(A==4){lines(pdf$x[2049:4096],X,lwd=2,lty=2,col='green3')}
if(A==5){lines(pdf$x[2049:4096],X)}}
```

```
#Calculating the Bayesian effect of 477 years of no activity conditioning; simple MC.
MC_Posterior=function(M,VT,Vmu,VI0){Vs=numeric(M); for(j in 1:M){while(Vs[j]<=0){S=0; mu=rtrian(Vmu[1],Vmu[2],Vmu[3])
T=rtrian(VT[1],VT[2],VT[3]); I0=rtrian(VI0[1],VI0[2],VI0[3]); Vs[j]=IVerupt01hawk(T,mu,I0,1)}; if(j/10==floor(j/10)){print(j); flush.console()}}
Vs1=quantile(Vs,0.05); Vsm=mean(Vs); Vs2=quantile(Vs,0.95); plot_kdeT(Vs); print(round(c(Vs1,Vsm,Vs2))); flush.console(); return(Vs)}
```

```
#Nested MC on NxM samples calculates next eruption time mean and percentiles curves and plot them; in addition it calculates the
#percentiles of such curves and their uncertainty bounds; parameter A imposes the color; if f=1 plots only the mean curve, with line type t.
Perc_IVepoHawk=function(N,M,VT,Vmu,VI0,A=0,f=0,t=1){          lines(X,outp[,1],lwd=1,lty=2,col=B)
Vs=numeric(N); VsM=numeric(M)                                lines(X,outp[,2],lwd=1,lty=2,col=B)
for(j in 1:M){mu=0; T=0; I0=0                                lines(X,outp[,4],lwd=1,lty=1,col=B)
while(mu<=0){mu=rtrian(Vmu[1],Vmu[2],Vmu[3])}                q05=q95=numeric(M)
while(T<=0){T=rtrian(VT[1],VT[2],VT[3])}                     for(i in 1:M){q05[i]=qexp(0.05,VsM[i])
while(I0<=0){I0=rtrian(VI0[1],VI0[2],VI0[3])}; VsM[j]=I0     q95[i]=qexp(0.95,VsM[i])}
for(i in 1:N){Vs[i]=IVerupt01hawk(T,mu,I0,2)}               Vs1=quantile(VsM,0.95); Vsm=mean(VsM)
VsM[j]=fitdistr(Vs,'exponential')$estimate                  Vs2=quantile(VsM,0.05)
if(j/10==floor(j/10)){print(j); flush.console()}}           q1_1=quantile(q05,0.05); q1_m=mean(q05)
X=(seq(1,10001)-1)/10000*1250; Cv=matrix(0,10001,M)         q1_2=quantile(q05,0.95)
for(j in 1:M){Cv[,j]=dexp(X,VsM[j])}                        q2_1=quantile(q95,0.05); q2_m=mean(q95)
outp=seek_quantile(Cv,10001)                                q2_2=quantile(q95,0.95)
if(A==0){B='black'}; if(A==1){B='red'}; if(A==2){B='darkviolet'}  print(round(c(q1_1,1/Vs1,q2_1))); flush.console()
if(A==3){B='blue'}; if(A==4){B='green3'}                    print(round(c(q1_m,1/Vsm,q2_m))); flush.console()
if(f){lines(X,outp[,4],lwd=2,lty=t,col=B); return(NULL)}    print(round(c(q1_2,1/Vs2,q2_2))); flush.console()}
```

6.4. Principal Notation

The following is the list of the variables adopted in the thesis. Each variable is briefly described and if possible we report also the Definition (D) in which it is first introduced. The list is organized in four paragraphs concerning the main parts of the thesis. We focus on Chapters 1-4, and we do not include the variables that are specific of the more technical Chapters 5 and 6.

Doubly stochastic modelling

$(\Omega, \mathcal{F}, P)$ - sample space with the associate σ-algebra and probability measure. An arbitrary element of Ω is called ω. - D1.1.

$(E, \mathcal{E})$ - epistemic space with the associate σ-algebra; an arbitrary element from E is called e. - D1.1.

$(W, \mathcal{W})$ - physical space with the associate σ-algebra. - D1.1.

ξ - random variable from $(\Omega, \mathcal{F}, P)$ to $(E, \mathcal{E})$ sampling epistemic assumptions. - D1.1.

$\chi(e, \cdot)$ - random variable from $(\Omega, \mathcal{F}, P)$ to $(W, \mathcal{W})$ sampling the physical observables, conditional on e. - D1.1.

η - probability distribution on $(E, \mathcal{E})$, image of P under ξ. - D1.1.

$M(e)$ - probability distribution on $(W, \mathcal{W})$, image of P under $\chi(e, \cdot)$, conditional on e. - D1.1.

Vent opening probability maps

$A \subseteq \mathbb{R}^2$ - domain representing the area of the volcanic system, associated with its Borel σ-algebra $\mathcal{B}(A)$. - D1.2.

X - random variable from $(\Omega, \mathcal{F}, P)$ to $(A, \mathcal{B}(A))$ representing the location of the next eruptive vent. An arbitrary vent location in A is called x. - D1.2.

μ_X - probability distribution on $(A, \mathcal{B}(A))$, image of P under X. It is the probability map of vent opening. - D1.2.

π^1 - measurable function from $(W, \mathcal{W}, M)$ to $(A, \mathcal{B}(A))$, representing the projection of the physical space onto the space of the vent opening location. - D1.3.

$\check{X}(e, \cdot)$ - random variable from $(\Omega, \mathcal{F}, P)$ to $(A, \mathcal{B}(A))$ representing the location of the next eruptive vent conditional on e. - D1.3.

$\mu_{\check{X}}(e)$ - probability distribution on $(A, \mathcal{B}(A))$, image of $M(e)$ under π^1, or of P under $\check{X}(e, \cdot)$. It is the probability map of vent opening conditional on e. - D1.3.

$(X_i)_{i=1,...,d}$ - random variables from $(\Omega, \mathcal{F}, P)$ to $(A, \mathcal{B}(A))$ representing vent locations based on the information of single volcanologic features. Their distributions of $(A, \mathcal{B}(A))$ are called $(\mu_i)_{i=1,...,d}$. - D2.4.

$(\alpha_i)_{i=1,...,d}$ - random variables from $(E, \mathcal{E}, \eta)$ to $([0, 1], \mathcal{B}(0, 1))$ representing the linear weights of the maps. - D2.4.

$\beta = (\beta_j)_{j=1,...,d'}$ - random variables defined on $(E, \mathcal{E}, \eta)$ representing the DM responses to the elicitation questionnaire concerning vent opening. f is a measurable function that permits to calculate $(\alpha_i)_{i=1,...,d}$ from $\beta = (\beta_j)_{j=1,...,d'}$. - D2.5.

$\mathcal{V} = (w_i)_{i=1,...,n}$ - discrete set of all the eruptive events considered. - D2.6.

$(D_i)_{i=1,...,n}$ - ellipses representing the enlarged locations of the eruptions. For each i, ζ_i is a uniform probability measure supported on D_i. - D2.6.

$(A_l)_{l=1,...,N}$ - finite partition of the spatial domain A. - D2.7.

Appendix A of Chapter 2 relies on the previous notation.

Appendix B of Chapter 2, re-defines some previously adopted symbols with different meanings (not reported here) valid only in that section.

Pyroclastic density current invasion maps
Y - real positive random variable from $(\Omega, \mathcal{F}, P)$ representing the location of the area invaded by PDCs during the next explosive eruption. An arbitrary areal size is called y. - D1.4.

ν_Y - probability distribution on $(\mathbb{R}_+, \mathcal{B}(\mathbb{R}_+))$, image of P under Y. It is the distribution of PDC invaded areas. - D1.4.

π^2 - measurable function from $(W, \mathcal{W}, M)$ to $(\mathbb{R}_+, \mathcal{B}(\mathbb{R}_+))$, representing the projection of the physical space onto the eruptive scale space. - D1.5.

$\check{Y}(e, \cdot)$ - random variable from $(\Omega, \mathcal{F}, P)$ to $(\mathbb{R}_+, \mathcal{B}(\mathbb{R}_+))$ representing the area invaded by the next PDC, conditional on e. - D1.5.

$\nu_{\check{Y}}(e)$ - probability distribution on $(\mathbb{R}_+, \mathcal{B}(\mathbb{R}_+))$, image of $M(e)$ under π^2, or of P under $\check{Y}(e, \cdot)$. It is the probability distribution of PDC invaded area conditional on e. - D1.5.

$B \subseteq \mathbb{R}^2$ - domain representing an enlarged zone possibly affected by PDC hazard, associated with its Borel σ-algebra $\mathcal{B}(A)$. An arbitrary element of B is called z. - D1.6.

F - function from $A \times \mathbb{R}_+$ to the Borel subsets of B, representing the set invaded by a PDC propagating from a vent x with a scale y. - D1.6.

p - measurable function from $(B, \mathcal{B}(B))$ to $([0, 1], \mathcal{B}([0, 1])$ coinciding with $E[1_{F(X,Y)}]$; represents the probability of each point of B to be reached by the next PDC. - D1.7.

$\check{p}(z)$ - random variable from $(E, \mathcal{E}, \eta)$ to $([0, 1], \mathcal{B}([0, 1])$ representing the probability of each point of B to be reached by the next PDC, conditional on e. - D1.7.

(γ_1, γ_2) - real positive random variables defined on $(E, \mathcal{E}, \eta)$ representing mean and standard deviation of the distribution of PDC invaded areal sizes. - D3.6.

$\tilde{\beta} = (\beta_j)_{j \in d'+1,...,d'+q}$ - random variables defined on $(E, \mathcal{E}, \eta)$ representing the DM responses to the elicitation questionnaire concerning PDC assessment. g is a measurable function that permits to calculate γ from $\tilde{\beta}$. - D3.6.

l_{max} - real positive arbitrary number representing the maximum run-out of a PDC neglecting topography. - D3.7.

K - real positive measurable function defined on $[0, diam(B)] \times \mathbb{R}_+$ with their Borel σ-algebra, representing the kinetic energy as a function of the distance r from the vent, and of l_{max}. - D3.7.

U - real positive measurable function defined on $(A, \mathcal{B}(A))$, representing the potential energy associated with the topography, as a function of the location. - D3.7.

$\tilde{F}$ - function from $A \times \mathbb{R}_+$ to the Borel subsets of B, representing the set invaded by a PDC propagating from a vent x with a scale y, based on the kinetic energy comparison, without shading. - D3.7.

$\sigma(\cdot)$ - set function from $A \times \mathcal{P}(B)$ on $\mathcal{P}(B)$ representing the shading of the area invaded. - D3.8.

R - real positive measurable function defined on $A \times \mathbb{R}_+$ representing the parameter l_{max} as a function of the vent location x and of the areal size y. - D3.8.

Appendix A of Chapter 3, re-defines some previously adopted symbols with different meanings (not reported here) valid only in that section.

Appendix B of Chapter 3, includes the following notation for the physical parameters.

$l(t)$ - position of the front of the current as a function of time.

$u(t)$ - velocity of the front of the current as a function of time.

$h(t)$ - height of the current as a function of time.

$\phi(t)$ - volume fraction of particles as a function of time.

V - volume of the current per radiant.

Fr - Froude number.

w_s - sedimentation velocity.

g_p - reduced gravity of the particles.

H - potential energy required for overcoming an obstacle.

C - physical constant summarizing all the parameters governing the box model approximation.

Appendix C of Chapter 3 mostly relies on the previous notation, and introduces a few new variables not reported here.

Time-space model for the next eruption

τ - random variable from $(E, \mathcal{E}, \eta)$ to the space $\mathcal{S}(n)$ of the permutations of $\{1, \ldots, n\}$, representing the random time sequence of past eruptions. - D1.8.

$(v_j)_{j=1,\ldots,n}$ - set of all the eruptive events considered, with an ordering imposed by τ. - D1.8.

$(t_j)_{j=1,\ldots,n}$ - vector of real random variables from $(E, \mathcal{E}, \eta)$ to $\mathbb{R}^n_+$, each t_j representing the time of eruptive event v_j. - D1.8.

V_j - random variable from $(E, \mathcal{E}, \eta)$ to $(A, \mathcal{B}(A))$ representing the location of the eruption v_j. - D1.8.

Θ_l - random set of random variables representing the times of each eruption v_j that occurred in the zone A_l. - D1.8.

$Z = (Z^l)_{l=1,\ldots,N}$ - doubly stochastic multivariate Hawkes process on $(\Omega, \mathcal{F}, P)$ representing the times and locations of eruptions. - D1.9.

φ - functional from E to the space of continuous decreasing functions on $\mathbb{R}_+$, representing the random diminishing of self interaction for the process Z. - D1.9.

λ_0^l - real positive random variable from $(E, \mathcal{E})$ representing the base rate of the process Z^l. - D1.9.

π^3 - measurable function from $(W, \mathcal{W}, M)$ to the space of l-dimensional counting measures, representing the projection of the physical space onto the set of next eruptions times in each of the caldera zones. - D1.10.

$\check{Z}(e, \cdot)$ - point process defined on $(\Omega, \mathcal{F}, P)$ representing the set of next eruptions times, conditional on e. - D1.10.

Z_{mn} - Cox-Hawkes process starting from a situation without excitement except for the residual additional intensity from Monte Nuovo, after $t_0 = 477$ years. - D1.11.

Z^* - positive random variable on $(\Omega, \mathcal{F}, P)$ representing the remaining time before the next eruption at Campi Flegrei. - D1.11.

ϱ_{Z^*} - probability distribution on $(\mathbb{R}_+, \mathcal{B}(\mathbb{R}_+))$, image of P under Z^*. It is the distribution of next eruption time from the present (year 2015). - D1.11.

$\check{Z}_{mn}$ - Cox-Hawkes process starting from a situation without excitement except for the residual additional intensity from Monte Nuovo, after $t_0 = 477$ years, conditional on e. - D1.12.

$\check{Z}^*(e, \cdot)$ - random variable from $(\Omega, \mathcal{F}, P)$ to $(\mathbb{R}_+, \mathcal{B}(\mathbb{R}_+))$ representing the remaining time before the next eruption at Campi Flegrei, conditional on e. - D1.12.

$\varrho_{\check{Z}^*}(e)$ - probability distribution on $(\mathbb{R}_+, \mathcal{B}(\mathbb{R}_+))$, image of P under $\check{Z}^*(e, \cdot)$. It is the probability distribution of next eruption time conditional on e. - D1.12.

k, h - real positive random variables on the space $(E, \mathcal{E}, \eta)$ representing the parameters of the self-interaction function of an Hawkes process. - D4.4.

L - likelihood function of a time-space record. L^l is the likelihood of the sub-record of zone A_l. - D4.5.

$\tilde{Z}^3$ - counting process without base rate and including only the residual self-excitement from Monte Nuovo event after $t_0 = 477$ years. - D4.9.

Q_{mn} - probability of producing an offspring after $t_0 = 477$ years of quiescence from the first event, in absence of other previous excitements. - D4.9.

L_{mn} - likelihood of passing $t_0 = 477$ years without other events after one eruption, in absence of other previous excitements. - D4.9.

T - real parameter representing the duration of self-excitement: the integrated additional intensity on the times above T is 5% of the total.

μ - real parameter representing the mean of the offspring points from a single ancestor (not including offspring of an offspring).

Appendix A of Chapter 4 mostly relies on the previous notation, and introduces some new variables not reported here.

References

[1] V. ACOCELLA, *Understanding caldera structure and development: an overview of analogue models compared to natural calderas*, Earth-Sci. Rev. **85** (2007), 125–160.

[2] V. ACOCELLA, *Activating and reactivating pairs of nested collapses during caldera-forming eruptions in Campi Flegrei* (Italy), Geophys. Res. Lett. **35** (2008).

[3] I. ALBERICO, L. LIRER, P. PETROSINO and R. SCANDONE, *A methodology for the evaluation of long-term volcanic risk from pyroclastic flows in Campi Flegrei* (Italy), J. Volcanol. Geoth. Res. **116** (2002), 63–78.

[4] I. ALBERICO, P. PETROSINO and L. LIRER, *Volcanic hazard and risk assessment*, In: "A Multi-source Volcanic Area", the example of Napoli city (Southern Italy), Nat. Hazard Earth Sys. **11** (2011), 1057–1070.

[5] D. ANDRONICO and L. LODATO, *Effusive activity at Mount Etna volcano*, (Italy) During the 20th century: a contribution to volcanic hazard assessment, Nat. Hazards **36** (2006), 407–443.

[6] S. ARRIGHI, C. PRINCIPE and M. ROSI, *Violent strombolian and subplinian eruptions at Vesuvius during post-1631 activity*, Bull. Volcanol. **63** (2001), 126–150.

[7] W. P. ASPINALL, *Structured elicitation of expert judgment for probabilistic hazard and risk assessment in volcanic eruptions*, In: "Statistics in Volcanology", H. M. Mader *et al.* (eds.), Geological Society of London on behalf of IAVCEI, 2006, 15–30.

[8] W. P. ASPINALL, *A route to a more tractable expert advice*, Nature **463** (2010), 294–295.

[9] W. P. ASPINALL and R. BLONG, *Volcanic Risk Assessment*, In: "The Encyclopedia of Volcanoes" (second edition), H. Sigurdsson *et al.* (eds.), Chapter 70, 2015, 1215–1231.

[10] T. BAI and D. D. POLLARD, *Fracture spacing in layered rocks: a new explanation based on the stress transition*, J. Struct. Geol. **22** (2000), 43–57.

[11] J. BARCLAY, K. HAYNES, B. HOUGHTON and D. JOHNSTON, *Social Processes and Volcanic Risk Reduction*, In: "The Encyclopedia of Volcanoes" (second edition), H. Sigurdsson *et al.* (eds.), Chapter 70, 2015 1203–1214.

[12] S. BARTOLINI, A. CAPPELLO, J. MARTÌ and C. DEL NEGRO, *Q-VAST: A new Quantum GIS plug-in for estimating volcanic susceptibility*, Nat. Hazard Earth Sys. **13** (2013), 3031–3042.

[13] P. J. BAXTER, R. BOYD, P. COLE, A. NERI, R. SPENCE and G. ZUCCARO, The impacts of pyroclastic surges on buildings at the eruption of the Soufrière Hills Volcano, Montserrat, Bull. Volcanol. **67** (2005), 292–313.

[14] P. J. BAXTER, W. P. ASPINALL, A. NERI, G. ZUCCARO, R. S. J. SPENCE, R. CIONI and G. WOO, *Emergency planning and mitigation at Vesuvius: A new evidence-based approach*, J. Volcanol. Geoth. Res. **178** (2008), 454–473.

[15] M. S. BEBBINGTON and S. J. CRONIN, *Spatio-temporal hazard estimation in the Auckland Volcanic Field*, New Zealand, with a new vent-order model, Bull. Volcanol. **73** (2011), 55–72.

[16] G. BERRINO, G. CORRADO, G. LUONGO and B.TORO, *Ground deformation and gravity changes accompanying the 1982 Pozzuoli uplift*, Bull. Volcanol. **47** (1984), 187–200.

[17] A. BEVILACQUA, R. ISAIA, A. NERI, S. VITALE, W. P. ASPINALL, M. BISSON, F. FLANDOLI, P. J. BAXTER, A. BERTAGNINI, T. ESPOSTI ONGARO, E. IANNUZZI, M. PISTOLESI and M. ROSI, Quantifying volcanic hazard at Campi Flegrei caldera (Italy) with uncertainty assessment, I. Vent opening maps, J. Geophys. Res. **120** (2015), 2309–2329.

[18] M. BISSON, A. FORNACIAI and F. MAZZARINI, *SITOGEO: A geographic database used for GIS applications*, Il Nuovo Cimento C – Note Brevi, **30C** (2007).

[19] C. BONADONNA, A. COSTA, A. FOLCH and T. KOYAGUCHI, *Tephra Dispersal and Sedimentation*, In: "The Encyclopedia of Volcanoes" (second edition), H. Sigurdsson *et al.* (eds.), Chapter 33, 2015, 587–597.

[20] M. J. BRANNEY and B. P KOKELAAR, *Pyroclastic density currents and the sedimentation of ignimbrites*, Geo. Soc. Mem. **27** (2002), 8 pp.

[21] M. J. BRANNEY and V. ACOCELLA, *Calderas*, In: "The Encyclopedia of Volcanoes" (second edition), H. Sigurdsson *et al.* (eds.), Chapter 16, 2015, 299–315.

[22] P. BRÉMAUD and L. MASSOULIÉ, *Stability of Nonlinear Hawkes Processes*, Ann. Probab. **24** (1996), 1563–1588.

[23] R. J. BROWN and G. D. ANDREWS, *Deposits of pyroclastic density currents*, In: "The Encyclopedia of Volcanoes" (second edition), H. Sigurdsson *et al.* (eds.), Chapter 36, 2015, 631–648.

[24] E. CALAIS, *et al.*, *Strain accommodation by slow slip and dyking in a youthful continental rift*, East Africa, Nature **456** (2008), 783–788.

[25] E. S. CALDER, P. D. COLE, W. B. DADE, T. H. DRUITT, R. P. HOBLITT, H. E. HUPPERT, L. RITCHIE, R. S. J. SPARKS and S. R. YOUNG, *Mobility of pyroclastic flows and surges at the Soufrière Hills volcano*, Montserrat, Geophys. Res. Lett., **26** (1999), 537–540.

[26] E. S. CALDER, Y. LAVALLE, J. E. KENDRICK and M. BERNSTEIN, *Lava dome eruptions*, In: "The Encyclopedia of Volcanoes" (second edition), H. Sigurdsson *et al.* (eds.), Chapter 18, 2015, 343–362.

[27] A. CAPPELLO, M. NERI, V. ACOCELLA, G. GALLO, A. VICARI and C. DEL NEGRO, *Spatial vent opening probability map of Etna volcano* (Sicily, Italy), Bull. Volcanol. **74** (2012), 2084–2094.

[28] P. CAPUANO, G. RUSSO, L. CIVETTA, G. ORSI, M. D'ANTONIO and R. MORETTI, *The active portion of the Campi Flegrei caldera structure imaged by 3-D inversion of gravity data*, Geochem. Geophy. Geosy. **14** (2013), 4681–4697.

[29] S. CAREY and M. BURSIK, *Volcanic plumes*, In: "The Encyclopedia of Volcanoes" (second edition), H. Sigurdsson *et al.* (eds.), Chapter 32, 2015, 571–585.

[30] K. V. CASHMAN and B. SCHEU, *Magmatic fragmentation*, In: "The Encyclopedia of Volcanoes" (second edition), H. Sigurdsson *et al.* (eds.), Chapter 25, 2015, 459–471.

[31] N. CHAPMAN, M. APTED, W. ASPINALL, K. BERRYMAN, M. CLOOS, C. B. CONNOR, L. CONNOR, O. JAQUET, K. KIYOSUGI, E. SCOURSE, S. SPARKS, M. STIRLING, L. WALLACE and J. GOTO, *TOPAZ Project: Long-term tectonic hazard to geological repositories*, Nuclear Waste Management Organization of Japan (NUMO) Report (2012), 87 pp.

[32] S. J. CHARBONNIER and R. GERTISSER, *Numerical simulations of block-and-ash flows using the Titan2D flow model: examples*

from the 2006 eruption of Merapi Volcano, Java, Indonesia, Bull. Volcanol. **71** (2009), 953–959.

[33] S. J. CHARBONNIER, A. GERMA, C. B. CONNOR, R. GER-TISSER, K. PREECE, J. C. KOMOROWSKI, F. LAVIGNE, T. DIXON, L. CONNOR, *Evaluation of the impact of the 2010 pyroclastic density currents at Merapi volcano from high-resolution satellite imagery, field investigations and numerical simulations*, J. Volcanol. Geoth. Res. **261** (2013), 295–315.

[34] G. CHIODINI, S. CALIRO, P. DE MARTINO, R. AVINO and F. GHEPARDI, *Early signals of new volcanic unrest at Campi Flegrei caldera? Insights from geochemical data and physical simulations*, Geology **40** (2012), 943–946.

[35] R. CIONI, A. BERTAGNINI, R. SANTACROCE and D. ANDRONICO, *Explosive activity and eruption scenarios at Somma-Vesuvius (Italy): Towards a new classification scheme*, J. Volcanol. Geoth. Res. **178** (2008), 331–346.

[36] P. D. COLE, A. NERI and P. J. BAXTER, *Hazards from pyroclastic density currents*, In: "The Encyclopedia of Volcanoes" (second edition), H. Sigurdsson *et al.* (eds.), Chapter 54, 2015, 943–956.

[37] L. CONNOR, C. B. CONNOR, K. MELIKSETIAN and I. SAVOV, *Probabilistic approach to modeling lava flow inundation: a lava flow hazard assessment for a nuclear facility in Armenia*, J. Appl. Volcanol. **1** (3) (2012).

[38] C. B. CONNOR and B. E. HILL, *Three nonhomogenous Poisson models for the probability of basaltic volcanism: application to the Yucca Mountain region*, Nevada, J. Geophys. Res. **100** (1995), 107–10, 125.

[39] C. B. CONNOR, J. A. STAMATAKOS, D. A. FERRILL, B. E. HILL, G. I. OFOEGBU, F. M. CONWEY, B. SAGAR and J. TRAPP, *Geologic factors controlling patterns of small-volume basaltic volcanism: application to a volcanic hazards assessment at Yucca Mountain*, Nevada, J. Geophys. Res. **105** (2000), 417–432.

[40] C. B. CONNOR, M. S. BEBBINGTON and W. MARZOCCHI, *Probabilistic volcanic Hazard assessment*, In: "The Encyclopedia of Volcanoes" (second edition), H. Sigurdsson *et al.* (eds.), Chapter 51, 2015, 897–910.

[41] R. M. COOKE, "Experts in Uncertainty: Opinion and Subjective Probability in Science", Oxford Univ. Press, New York, 1991, 336 pp.

[42] R. M. COOKE and L. H. J. GOOSSENS, *TU delft expert judgement data base*, Reliab. Eng. Syst. Safe. **93** (2008), 657–674.

[43] R. M. COOKE, S. EL SAADANY and X. HUANG, *On the performance of social network and likelihood-based expert weighting schemes*, Reliab. Eng. Syst. Safe. **93** (2008), 745–756.

[44] D. R. COX and V. ISHAM, "Point Processes", Chapman and Hall, London and New York, 1980, 188 pp.

[45] W. B. DADE and H. E. HUPPERT, *Emplacement of the Taupo ignimbrite by a dilute turbulent flow*, Nature **381** (1996), 509–512.

[46] D. J. DALEY and D. VERE JONES "An Introduction to the Theory of Point Processes", (2005 Vol. I, 2008 Vol II), Springer, 469 + 573 pp.

[47] S. J. DAY, *Volcanic tsunamis*, In: "The Encyclopedia of Volcanoes" (second edition), H. Sigurdsson *et al.* (eds.), Chapter 58, 2015, 993–1009.

[48] A. L. DEINO, G. ORSI, S. DE VITA and M. PIOCHI, *The age of the Neapolitan Yellow Tuff caldera-forming eruption* (Campi Flegrei caldera, Italy) assessed by 40Ar/39Ar dating method, J. Volcanol. Geoth. Res. **133** (2004), 157–170.

[49] S. DELATTRE, N. FOURNIER and M. HOFFMANN, *Hawkes processes on large networks*, Ann. Appl. Probab. **26** (2015).

[50] C. DEL GAUDIO, I. AQUINO, G. P. RICCIARDI, C. RICCO and R. SCANDONE, *Unrest episodes at Campi Flegrei: a reconstruction of vertical ground movements during 1905-2009*, J. Volcanol. Geoth. Res. **195** (2010), 48–56.

[51] P. DELLINO, R. ISAIA and M. VENERUSO, *Turbulent boundary layer shear flows as an approximation of base surges at Campi Flegrei* (Southern Italy), J. Volcanol. Geoth. Res. **133** (2004), 211–228.

[52] S. DE SILVA and J. M. LINDSAY, *Primary volcanic landforms*, In: "The Encyclopedia of Volcanoes" (second edition), H. Sigurdsson *et al.* (eds.), Chapter 15, (2015), 273-297.

[53] S. DE VITA, G. ORSI, L. CIVETTA, A. CARANDENTE, M. D'ANTONIO, A. DEINO, T. DI CESARE, M. A. DI VITO, R. V. FISHER, R. ISAIA, E. MAROTTA, A. NECCO, M. ORT, L. PAPPALARDO, M. PIOCHI and J. SOUTHON, *The Agnano-Monte Spina eruption (4100 years BP) in the restless Campi Flegrei caldera* (Italy), J. Volcanol. Geoth. Res. **91** (1999), 269–301.

[54] V. DI RENZO, I. ARIENZO, L. CIVETTA, M. D'ANTONIO, S. TONARINI, M. A. DI VITO and G. ORSI, *The magmatic feeding system of of the Campi Flegrei caldera: architecture and temporal evolution*, Chem. Geol. **281** (2011), 227–241.

[55] B. DE VIVO, G. ROLANDI, P. B. GANS, A. CALVERT, W. A. BOHRSON, F. J. SPERA and A. E. BELKIN, *New constraints on*

the pyroclastic eruption history of the Campanian volcanic plain (Italy), Miner. Petrol. **73** (2001), 47–65.

[56] M. A. DI VITO, L. LIRER, G. MASTROLORENZO and G. ROLANDI, *The 1538 Monte Nuovo eruption (Campi Flegrei, Italy)*, Bull. Volcanol. **49** (1987), 608–615.

[57] M. A. DI VITO, R. ISAIA, G. ORSI, J. SOUTHON, S. DE VITA, M. D'ANTONIO, L. PAPPALARDO and M. PIOCHI, *Volcanism and deformation since 12,000 years at the Campi Flegrei caldera* (Italy), J. Volcanol. Geoth. Res. **91** (1999), 221–246.

[58] C. D'ORIANO, E. POGGIANTI, A. BERTAGNINI, R. CIONI, P. LANDI, M. POLACCI and M. ROSI, *Changes in eruptive style during the AD 1538 Monte Nuovo eruption (Phlegrean Fields, Italy): the role of syn-eruptive crystallization*, Bull. Volcanol. **67** (2005), 601–621

[59] T. H. DRUITT, *Pyroclastic density currents*, In: "The Physics of Explosive Volcanic Eruptions", J. S. Gilbert and R. S. J. Sparks (eds.), Geological Society Special Publications no. 145, Geological Society, London, 21, 1998, 145–182.

[60] J. DUFEK and G. W. BERGANTZ, *Suspended-load and bed-load transport of particle-laden gravity currents: the role of particle-bed interaction*, Theor. Comp. Fluid. Dyn. **21** (2007), 119–145.

[61] J. DUFEK, T. ESPOSTI ONGARO and O. ROCHE, *Pyroclastic density currents: processes and models*, In: "The Encyclopedia of Volcanoes" (second edition), H. Sigurdsson *et al.* (eds.), Chapter 35, 2015, 617–629.

[62] T. DUONG, *Ks: kernel density estimations and kernel discriminant analysis for multivariate data in R*, J. Stat. Softw. **21** (2007), 1–16.

[63] J. J. DVORAK and P. GASPARINI, *History of earthquakes and vertical ground movement in Campi Flegrei caldera, southern Italy: comparison of precursory events to the A.D. 1538 eruption of Monte Nuovo and of activity since 1968*, J. Volcanol. Geoth. Res. **48** (1991), 77–92.

[64] T. ESPOSTI ONGARO, C. CAVAZZONI, G. ERBACCI, A. NERI and M. V. SALVETTI, *A parallel multiphase flow code for the 3D simulation of volcanic explosive eruptions*, Parallel Comput. **33** (2007), 7–8, 541–560.

[65] T. ESPOSTI ONGARO, P. MARIANELLI, M. TODESCO, A. NERI, C. CAVAZZONI and G. ERBACCI, *Mappe tematiche, georeferenziate e digitali, delle principali azioni pericolose associate alle colate piroclastiche del Vesuvio e dei Campi Flegrei derivanti dalle nuove simulazioni 3D*, Prodotto D2.3.5, Progetto SPEED (in Italian) (2008a).

[66] T. ESPOSTI ONGARO, A. NERI, G. MENCONI, M. DE' MICHIE-
LI VITTURI, P. MARIANELLI, C. CAVAZZONI, G. ERBACCI and
P. J. BAXTER, *Transient 3D numerical simulations of column collapse and pyroclastic density current scenarios at Vesuvius*, J. Volcanol. Geoth. Res. **178** (2008b), 378–396.
[67] T. ESPOSTI ONGARO, C. WIDIWIJAYANTI, A. B. CLARKE, B.
VOIGHT and A. NERI, *Multiphase-flow numerical modeling of the
18 May 1980 lateral blast at Mount St. Helens*, USA, Geology **39**
(2011), 535–539.
[68] F. FLANDOLI, E. GIORGI, W. P. ASPINALL and A. NERI, *Comparison of a new expert elicitation model with the Classical Model,
equal weights and single experts, using a cross-validation technique*, Reliab. Eng. Syst. Safe. **96** (2011), 1292–1311.
[69] G. FLORIO, M. FEDI, F. CELLA and A. RAPOLLA, *The Campanian Plain and Phlegrean Fields: structural setting from potential
field data*, J. Volcanol. Geoth. Res. **91** (1999), 361–37.
[70] E. S. GAFFNEY, B. DAMJANAC and G. A. VALENTINE, *Localization of volcanic activity: 2. Effects of pre-existing structure*, Earth
Planet. Sc. Lett. **263** (2007), 323–338.
[71] B. GIACCIO, R. ISAIA, F. FEDELE, E. DI CANZIO, J. HOF-
FECKER, A. RONCHITELLI, A. SINITSYN, M. ANIKOVICH, S.
LISITSYN and V. POPOV, *The Campanian Ignimbrite and Codola
tephra layers: two temporal/stratigraphic markers for the Early
Upper Palaeolithic in southern Italy and eastern Europe*, J. Volcanol. Geoth. Res. **177** (2008), 208–226.
[72] C. GLADSTONE and A. W. WOODS, *On the application of box
models to particle-driven gravity currents*, J. Fluid. Mech. **416**
(2000), 187–195.
[73] H. GONNERMANN and B. TAISNE, *Magma Transport in Dikes*,
In: "The Encyclopedia of Volcanoes" (second edition), H. Sigurdsson *et al.* (eds.), Chapter 10, 2015, 215–224.
[74] C. E. GREGG, B. HOUGHTON and J. W. EWERT, *Volcano warn-
ing systems*, In: "The Encyclopedia of Volcanoes" (second edition), H. Sigurdsson *et al.* (eds.), Chapter 67, 2015, 1173–1185.
[75] A. GUDMUNDSSON, *Surface stresses associated with arrested
dykes in rift zones*, Bull. Volcanol. **65** (2003), 606–619.
[76] M. T. GUDMUNDSSON, *Hazards from Lahars and Jökulhlaups*,
In: "The Encyclopedia of Volcanoes" (second edition), H. Sigurdsson *et al.* (eds.), Chapter 34, 2015, 971–984.
[77] V. GUERRIERO, S. VITALE, S. CIARCIA and S. MAZZOLI, *Improved statistical multi-scale analysis of fractured reservoir ana-
logues*, Tectonophysics **504** (2011), 14–24.

[78] M. GUFFANTI and A. TUPPER, *Volcanic ash hazards and aviation risk*, In: "Volcanic Hazards, Risks and Disasters", P. Papale *et al.* (eds.), Chapter 4, 2014, 87–108.

[79] E. GUIDOBONI and C. CIUCCARELLI, *The Campi Flegrei caldera: historical revision and new data on seismic crises, bradyseisms, the Monte Nuovo eruption and ensuing earthquakes* (twelfth century 1582 AD), Bull. Volcanol. **73** (2011), 655–677.

[80] M. A. HALLWORTH, A. J. HOGG and H. E. HUPPERT, *Effects of external flow on compositional and particle gravity currents*, J. Fluid. Mech. **359** (1998), 109–142.

[81] T. C. HARRIS, A. J. HOGG and H. E. HUPPERT, *Polydisperse particle-driven gravity currents*, J. Fluid. Mech. **472** (2002), 333–371.

[82] A. J. L. HARRIS, *Basaltic lava flow hazard*, In: "Volcanic Hazards, Risks and Disasters", P. Papale *et al.* (eds.), Chapter 2, 2014, 17–46.

[83] A. J. L. HARRIS and S. ROWLAND, *Lava flows and rheology*, In: "The Encyclopedia of Volcanoes" (second edition), H. Sigurdsson *et al.* (eds.), Chapter 67, 2015, 321–342.

[84] B. HOUGHTON and R. J. CAREY, *Pyroclastic Fall deposits*, In: "The Encyclopedia of Volcanoes" (second edition), H. Sigurdsson *et al.* (eds.), Chapter 34, 2015, 599–616.

[85] K. J. HSU, *Catastrophic debris streams (sturzstroms) generated by rockfalls*, Geol. Soc. Am. Bull. **86** (1975), 129–140.

[86] H. E. HUPPERT and J. E. SIMPSON, *The slumping of gravity currents*, J. Fluid. Mech. **99** (1980), 785–799.

[87] J. JACOD and A. N. SHIRIAEV "Limit Theorems for Stochastic Processes", Springer, 2003.

[88] O. JAQUET, S. LOW, B. MARTINELLI, V. DIETRICH and D. GILBY, *Estimation of volcanic hazards based on cox stochastic processes*, Phys. Chem. Earth (A) **25** (2000), 571–579.

[89] O. JAQUET and R. CARNIEL, *Estimation of volcanic hazards using geostatistical models*, In: "Statistics in Volcanology", H. M. Mader *et al.* (eds.), Geological Society of London on behalf of IAVCEI, 2006, 89–104.

[90] O. JAQUET, C. B. CONNOR and L. CONNOR, *Probabilistic methodology for long-term assessment of volcanic hazards*, Nucl. Technol. **163** (2008), 180–189.

[91] G. JOLLY and S. DE LA CRUZ, *Volcanic crisis management*, In: "The Encyclopedia of Volcanoes" (second edition), H. Sigurdsson *et al.* (eds.), Chapter 68, 2015, 1187–1202.

[92] J. B. KADANE and L. J. WOLFSON, *Experiences in elicitation*, J. Roy. Stat. Soc. D-Stat. **47** (1998), 3–19.

[93] C. R. J. KILBURN, *Lava flow hazards and modeling*, In: "The Encyclopedia of Volcanoes" (second edition), H. Sigurdsson *et al.* (eds.), Chapter 55, 2015, 957–969.

[94] M. KLAWONN, B. F. HOUGHTON, D. A. SWANSON, S. A. FAGENTS, P. WESSEL and C. J. WOLFE, *From field data to volumes: constraining uncertainties in pyroclastic eruption parameters*, Bull. Volcanol. **76** (2014), 839–854.

[95] M. KYNN *The "heuristics and biases" in expert elicitation*, J. Roy. Stat. Soc. A Sta. **171** (2008), 239–264.

[96] R. ISAIA, M. D'ANTONIO, F. DELL'ERBA, M. DI VITO and G. ORSI, *The Astroni volcano: the only example of closely spaced eruptions in the same vent area during the recent history of Campi Flegrei caldera* (Italy), J. Volcanol. Geoth. Res. 133 (2004), 171–192.

[97] R. ISAIA, P. MARIANELLI and R. SBRANA *Caldera unrest prior to intense volcanism in Campi Flegrei* (Italy) *at 4.0 ka B.P.: implications for caldera dynamics and future eruptive scenarios*, Geophys. Res. Lett. **36** (2009), L21303.

[98] R. ISAIA, S. VITALE, M. G. DI GIUSEPPE, E. IANNUZZI, F. D. A. TRAMPARULO and A. TROIANO, *Stratigraphy, structure and volcano-tectonic evolution of Solfatara maar-diatreme* (Campi Flegrei, Italy), Geol. Soc. Am. Bull. **128** (2015).

[99] N. LE CORVEC, T. MENAND and J. LINDSAY, *Interaction of ascending magma with pre-existing crustal fractures in monogenetic basaltic volcanism: an experimental approach*, J. Geophys. Res. **118** (2013), 968–984.

[100] L. LIRER, P. PETROSINO and I. ALBERICO, *Hazard assessment at volcanic fields: the Campi Flegrei case history*, J. Volcanol. Geoth. Res. **112** (2001), 53–74.

[101] J. R. LISTER and R. C. KERR, *Fluid-mechanical models of crack propagation and their application to magma transport in dykes*, J. Geophys. Res. **96** (1991), 10049–10077.

[102] W. J. MALFAIT, R. SEIFERT, S. PETITGIRARD, J. P. PERRILLAT, M. MEZOUAR, T. OTA, E. NAKAMURA, P. LERCH and C. SANCHEZ-VALLE, *Supervolcano eruptions driven by melt buoyancy in large silicic magma chambers*, Nat. Geosci. **7** (2014), 122–125.

[103] B. MARSH, *Magma chambers*, In: "The Encyclopedia of Volcanoes" (second edition), H. Sigurdsson *et al.* (eds.), Chapter 8, 2015, 185–201.

[104] J. MARTÌ and A. FELPETO, *Methodology for the computation of volcanic susceptibility: an example for mafic and felsic eruptions on Tenerife* (Canary Islands), J. Volcanol. Geoth. Res. **195** (2010), 69–77.

[105] J. MARTÌ, V. PINEL, C. LÒPEZ, A. GEYER, R. ABELLA, M. TÀRRAGA, M. J. BLANCO, A. CASTRO and C. RODRÌGUEZ, *Causes and mechanisms of El Hierro submarine eruption (2011-2012)* (Canary Islands), J. Geophys. Res. **118** (2013), 1–17.

[106] W. MARZOCCHI and G. WOO, *Principles of volcanic risk metrics: Theory and the case study of Mount Vesuvius and Campi Flegrei*, Italy, J. Geophys. Res. **114** (2009), B03213.

[107] W. MARZOCCHI and M. S. BEBBINGTON, *Probabilistic eruption forecasting at short and long time scales*, Bull. Volcanol. **74** (2012), 1777-1805.

[108] L. G. MASTIN and D. D. POLLARD, *Surface deformation and shallow dike intrusion processes at Inyo Craters*, Long Valley, California, J. Geophys. Res. **93** (1988), 13221–13235.

[109] G. S. MATTIOLI, B. VOIGHT, A. T. LINDE, I. S. SACKS, P. WATTS, C. WIDIWIJAYANTI and D. WILLIAMS, *Unique and remarkable dilatometer measurements of pyroclastic flow-generated tsunamis*, Geology **35** (2007), 25–28.

[110] F. MAZZARINI, D. KEIR and I. ISOLA, *Spatial relationship between earthquakes and volcanic vents in the central-northern Main Ethiopian Rift*, J. Volcanol. Geoth. Res. **262** (2013a), 123–133.

[111] F. MAZZARINI, T. O. ROONEY and I. ISOLA, *The intimate relationship between strain and magmatism: a numerical treatment of clustered monogenetic fields in the Main Ethiopian Rift*, Tectonics **32** (2013b), 49–64.

[112] P. A. MOTHES and J. W. VALLANCE, *Lahars at cotopaxi and tungurahua volcanoes, Ecuador: highlights from stratigraphy and observational records and related downstream hazards*, In: "Volcanic Hazards, Risks and Disasters", P. Papale *et al.* (eds.), Chapter 6, 2014, 141–168.

[113] M. NAKAMURA, K. OTAKI and S. TAKEUCHI, *Permeability and pore-connectivity variation of pumices from a single pyroclastic flow eruption: Implications for partial fragmentation*, J. Volcanol. Geoth. Res. **176** (2008), 302–314.

[114] A. NERI, T. ESPOSTI ONGARO, G. MACEDONIO and D. GIDASPOW, *Multiparticle simulation of collapsing volcanic columns and pyroclastic flows*, J. Geophys. Res. **108** (2003), 148–227.

[115] A. NERI, W. P. ASPINALL, R. CIONI, A. BERTAGNINI, P. J. BAXTER, G. ZUCCARO, D. ANDRONICO, S. BARSOTTI, P. D. COLE, T. ESPOSTI ONGARO, T. K. HINCKS, G. MACEDONIO, P. PAPALE, M. ROSI, R. SANTACROCE and G. WOO, *Developing an event tree for probabilistic hazard and risk assessment at Vesuvius*, J. Volcanol. Geoth. Res. **178** (2008), 397–415.

[116] A. NERI, T. ESPOSTI ONGARO, B. VOIGHT and C. WIDIWI-JAYANTI, *Pyroclastic density current hazards and risk*, In: "Volcanic Hazards, Risks and Disasters", P. Papale *et al.* (eds.), Chapter 5, 2014, 109–140.

[117] A. NERI, A. BERTAGNINI, A. BEVILACQUA, M. BISSON, T. ESPOSTI ONGARO, F. FLANDOLI, R. ISAIA, M. PISTOLESI, M. ROSI and S. VITALE, *Mappe di pericolosità da flussi piroclas-tici ai Campi Flegrei*, Scientific Report to Commissione Nazionale Grandi Rischi, Dipartimento della Protezione Civile, (2014), 26 pp. (confidential).

[118] A. NERI, A. BEVILACQUA, T. ESPOSTI ONGARO, R. ISAIA, W. P. ASPINALL, M. BISSON, F. FLANDOLI, P. J. BAXTER, A. BERTAGNINI, E. IANNUZZI, S. ORSUCCI, M. PISTOLESI, M. ROSI and S. VITALE, *Quantifying volcanic hazard at Campi Fle-grei caldera* (Italy) with uncertainty assessment: II. Pyroclastic density current invasion maps, J. Geophys. Res. **120** (2015), 2330–2349.

[119] Y. NISHIMURA, M. NAKAGAWA, J. KUDUON and J. WUKAWA, *Timing and scale of tsunamis caused by the 1994 Rabaul eruption, East New Britain, Papua New Guinea*, In: "Tsunamis", Springer, 2005, 43–56.

[120] A. O'HAGAN, C. E. BUCK, A. DANESHKHAN, J. R. EISER, P. H. GARTHWAITE, D. J. JENKINSON, J. E. OAKLEY and T. RAKOW, "Uncertain Judgements: Eliciting Experts' Probabili-ties", John Wiley & Sons, England, 2006, 338 pp.

[121] G. ORSI, M. D'ANTONIO, S. DE VITA and G. GALLO, *The Neapolitan Yellow Tuff, a large-magnitude trachytic phreato-plinian eruption; eruptive dynamics, magma withdrawal and caldera collapse*, J. Volcanol. Geoth. Res. **53** (1992), 275–287.

[122] G. ORSI, M. A. DI VITO and R. ISAIA, *Volcanic hazard as-sessment at the restless Campi Flegrei caldera*, Bull. Volcanol. **66** (2004), 514–530.

[123] G. ORSI, M. A. DI VITO, J. SELVA and W. MARZOCCHI, *Long-term forecast of eruptive style and size at Campi Flegrei caldera* (Italy), Earth Planet. Sc. Lett. **287** (2009), 265–276.

[124] O. ORTEGA, R. MARRETT and E. LAUBACH, *Scale-independent approach to fracture intensity and average spacing measurement*, AAPG Bull. **90** (2006), 193–208.

[125] J. PALLISTER and S. R. MCNUTT, *Synthesis of volcano monitoring*, In: "The Encyclopedia of Volcanoes" (second edition), H. Sigurdsson *et al.* (eds.), Chapter 66, 2015, 1151–1171.

[126] D. D. POLLARD, P. T. DELANEY, W. A. DUFFIELD, E. T. ENDO and A. T. OKAMURA, *Surface deformation in volcanic rift zones*, Tectonophysics **94** (1983), 541–584.

[127] F. PRATA and B. ROSE, *Volcanic ash hazards to aviation*, In: "The Encyclopedia of Volcanoes" (second edition), H. Sigurdsson *et al.* (eds.), Chapter 52, 2015, 911–934.

[128] A. RITTMANN, *Sintesi Geologica dei Campi Flegrei*, Bo. Soc. Geol. Ital. **LXIX-II** (1950), 117–128.

[129] N. ROGERS, *The composition and origin of magmas*, In: "The Encyclopedia of Volcanoes" (second edition), H. Sigurdsson *et al.* (eds.), Chapter 4, 2015, 93–112.

[130] K. ROGGENSACK, S. N. WILLIAMS, S. J. SCHAEFER and R. A. J. PARNELL, *Volatiles from 1994 eruptions of Rabaul: understanding large caldera systems*, Science **273** (1996), 490–493.

[131] T. O. ROONEY, I. D. BASTOW and D. KEIR, *Insights into extensional processes during magma assisted rifting: evidence from aligned scoria cones and maars*, J. Volcanol. Geoth. Res. **201**, 83–96.

[132] M. ROSI, A. SBRANA and C. PRINCIPE, *The Phlegrean Fields: structural evolution, volcanic history and eruptive mechanisms*, J. Volcanol. Geoth. Res. **17** (1983), 273–288.

[133] M. ROSI and A. SBRANA (eds.), "Phlegraean Fields", Quaderni della Ricerca Scientifica, CNR, Roma, Vol. 114, 1987, 175 pp.

[134] S. ROSSANO, G. MASTROLORENZO and G. DE NATALE, *Numerical simulation of pyroclastic density currents on Campi Flegrei topography: a tool for statistical hazard estimation*, J. Volcanol. Geoth. Res. **132** (2004), 1–14.

[135] R DEVELOPMENT CORE TEAM, "*R*: A Language and Environment for Statistical Computing, *R* Foundation for Statistical Computing", Vienna, 2008, http://www.R-project.org.

[136] A. M. RUBIN and D. D. POLLARD, *Dike-induced faulting in rift zones of Iceland and Afar*, Geology **16** (1988), 413–417.

[137] G. SACCOROTTI, M. IGUCHI and A. AIUPPA, *In situ volcano monitoring: present and future*, In: "Volcanic Hazards, Risks and Disasters", P. Papale *et al.* (eds.), Chapter 7, 2014, 169–202.

[138] C. SCARPATI, A. PERROTTA, S. LEPORE and A. CALVERT, *Eruptive history of neapolitan volcanoes: constrains from 40Ar-39Ar dating*, Geol. Mag. **150** (2012), 412–425.

[139] R. SCANDONE, F. BELLUCCI, L. LIRER and G. ROLANDI, *The structure of the campanian plain and the activity of the neapolitan volcanoes* (Italy), J. Volcanol. Geoth. Res. **48** (1991), 1–31.

[140] A. SCHMIDT, *Volcanic gas and aerosol hazards from a future laki-type eruption in iceland*, In: "Volcanic Hazards, Risks and Disasters", P. Papale *et al.* (eds.), Chapter 15, 2014, 377–397.

[141] J. SELVA, G. ORSI, M. A. DI VITO, W. MARZOCCHI and L. SANDRI, *Probability hazard map for future vent opening at the Campi Flegrei caldera* (Italy), Bull. Volcanol. **74** (2012), 497–510.

[142] J. SELVA, W. MARZOCCHI, L. SANDRI and A. COSTA, *Operational short-term volcanic hazard analysis: methods and perspectives*, In: "Volcanic Hazards, Risks and Disasters", P. Papale *et al.* (eds.), Chapter 9, 2014, 233-259.

[143] H. SHINOHARA, *Excess degassing from volcanoes and its role on eruptive and intrusive activity*, Rev. Geophys. **46** (2008), RG4005.

[144] L. SIEBERT, E. COTTRELL, E. VENZKE and B. ANDREWS, *Eart's volcanoes and their eruptions: an overview*, In: "The Encyclopedia of Volcanoes" (second edition), H. Sigurdsson *et al.* (eds.), Chapter 12, 2015, 239-255.

[145] V. C. SMITH, R. ISAIA and N. J. G. PEARCE, *Tephrostratigraphy and glass compositions of post-15 ka Campi Flegrei eruptions: implications for eruption history and chronostratigraphic markers*, Quaternary Sci. Rev. **30** (2011), 3638–3660.

[146] R. S. J. SPARKS and H. SIGURDSSON, *Magma mixing: a mechanism for triggering acid explosive eruptions*, Nature **267** (1977), 315–318.

[147] R. S. J. SPARKS, M. I. BURSIK, S. N. CAREY, J. S. GILBERT, L. S. GLAZE, H. SIGURDSSON and A. W. WOODS, "Volcanic Plumes", John Wiley & Sons, England, 1997, 574 pp.

[148] E. T. SPILLER, M. J. BAYARRI, J. O. BERGER, E. S. CALDER, A. K. PATRA, E. B. PITMAN, R. L. WOLPERT, *Automatic emulator construction for geophysical mass flows*, J. Uncertainty Quantification, SIAM/ASA **2** (2014), 126–152.

[149] Y. TATSUMI and K. SUZUKI-KAMATA, *Cause and risk of catastrophic eruptions in the Japanese Archipelago*, P. Jpn. Acad. B-Phys. **90** (2014).

[150] M. TODESCO, A. NERI, T. ESPOSTI ONGARO, P. PAPALE and M. ROSI, *Pyroclastic flow dynamics and hazard in a caldera setting: Application to Phlegrean Fields* (Italy), Gcube **7** (2006), 11.

[151] G. A. VALENTINE, *Stratified flow in pyroclastic surges*, Bull. Volcanol. **49** (1987), 616–630.

[152] J. W. VALLANCE and R. M. IVERSON, *Lahars and their deposits*, In: "The Encyclopedia of Volcanoes" (second edition), H. Sigurdsson *et al.* (eds.), Chapter 37, 2015, 649–664.

[153] S. VITALE and R. ISAIA, *Fractures and faults in volcanic rocks* (Campi Flegrei, Southern Italy): insights into volcano-tectonic processes, Int. J. Earth Sci. **103** (2014), 801–819.

[154] G. WADGE and W. P. ASPINALL, *A review of volcanic hazard and risk assessments at the Soufriere Hills Volcano*, Montserrat from 1997 to 2011, In: "The eruption of Soufriere Hills Volcano", Montserrat from 2000 to 2010, G. Wadge, R. E. Robertson, B. Voight (eds.), Memoir. 39, Geological Society, London, 2014.

[155] J. D. L. WHITE, C. I. SCHIPPER and K. KANO, *Submarine explosive eruptions*, In: "The Encyclopedia of Volcanoes" (second edition), H. Sigurdsson *et al.* (eds.), Chapter 31, 2015, 553-569.

[156] G. WILLIAMS-JONES and H. RYMER, *Hazards of volcanic gases*, In: "The Encyclopedia of Volcanoes" (second edition), H. Sigurdsson *et al.* (eds.), Chapter 57, 2015, 985–992.

[157] T. M. WILSON, S. JENKINS and C. STEWART, *Impacts from volcanic ash fall*, In: "Volcanic Hazards, Risks and Disasters", P. Papale *et al.* (eds.), Chapter 3, 2014, 47–86.

[158] G. WOO, "The Mathematics of Natural Catastrophes", Imperial College Press, Singapore, 1999, 292 pp.

[159] G. WOO, *Cost-benefit analysis in volcanic risk*, In: "Volcanic Hazards, Risks and Disasters", P. Papale *et al.*, (eds.) Chapter 11, 2015, 289-300.

THESES

This series gathers a selection of outstanding Ph.D. theses defended at the Scuola Normale Superiore since 1992.

Published volumes

1. F. COSTANTINO, *Shadows and Branched Shadows of 3 and 4-Manifolds*, 2005. ISBN 88-7642-154-8

2. S. FRANCAVIGLIA, *Hyperbolicity Equations for Cusped 3-Manifolds and Volume-Rigidity of Representations*, 2005. ISBN 88-7642-167-x

3. E. SINIBALDI, *Implicit Preconditioned Numerical Schemes for the Simulation of Three-Dimensional Barotropic Flows*, 2007. ISBN 978-88-7642-310-9

4. F. SANTAMBROGIO, *Variational Problems in Transport Theory with Mass Concentration*, 2007. ISBN 978-88-7642-312-3

5. M. R. BAKHTIARI, *Quantum Gases in Quasi-One-Dimensional Arrays*, 2007. ISBN 978-88-7642-319-2

6. T. SERVI, *On the First-Order Theory of Real Exponentiation*, 2008. ISBN 978-88-7642-325-3

7. D. VITTONE, *Submanifolds in Carnot Groups*, 2008. ISBN 978-88-7642-327-7

8. A. FIGALLI, *Optimal Transportation and Action-Minimizing Measures*, 2008. ISBN 978-88-7642-330-7

9. A. SARACCO, *Extension Problems in Complex and CR-Geometry*, 2008. ISBN 978-88-7642-338-3

10. L. MANCA, *Kolmogorov Operators in Spaces of Continuous Functions and Equations for Measures*, 2008. ISBN 978-88-7642-336-9

11. M. LELLI, *Solution Structure and Solution Dynamics in Chiral Ytterbium(III) Complexes*, 2009. ISBN 978-88-7642-349-9
12. G. CRIPPA, *The Flow Associated to Weakly Differentiable Vector Fields*, 2009. ISBN 978-88-7642-340-6
13. F. CALLEGARO, *Cohomology of Finite and Affine Type Artin Groups over Abelian Representations*, 2009. ISBN 978-88-7642-345-1
14. G. DELLA SALA, *Geometric Properties of Non-compact CR Manifolds*, 2009. ISBN 978-88-7642-348-2
15. P. BOITO, *Structured Matrix Based Methods for Approximate Polynomial GCD*, 2011. ISBN: 978-88-7642-380-2; e-ISBN: 978-88-7642-381-9
16. F. POLONI, *Algorithms for Quadratic Matrix and Vector Equations*, 2011. ISBN: 978-88-7642-383-3; e-ISBN: 978-88-7642-384-0
17. G. DE PHILIPPIS, *Regularity of Optimal Transport Maps and Applications*, 2013. ISBN: 978-88-7642-456-4; e-ISBN: 978-88-7642-458-8
18. G. PETRUCCIANI, *The Search for the Higgs Boson at CMS*, 2013. ISBN: 978-88-7642-481-6; e-ISBN: 978-88-7642-482-3
19. B. VELICHKOV, *Existence and Regularity Results for Some Shape Optimization Problems*, 2015. ISBN: 978-88-7642-526-4; e-ISBN: 978-88-7642-527-1
20. M. RUGGIERO, *Rigid Germs, the Valuative Tree, and Applications to Kato Varieties*, 2015. ISBN: 978-88-7642-558-5 e-ISBN: 978-88-7642-559-2
21. A. BEVILACQUA, *Doubly Stochastic Models for Volcanic Hazard Assessment at Campi Flegrei Caldera*, 2016. ISBN: 978-88-7642-556-1 e-ISBN: 978-88-7642-577-6

Volumes published earlier

H. Y. FUJITA, *Equations de Navier-Stokes stochastiques non homogènes et applications*, 1992.

G. GAMBERINI, *The minimal supersymmetric standard model and its phenomenological implications*, 1993. ISBN 978-88-7642-274-4

C. DE FABRITIIS, *Actions of Holomorphic Maps on Spaces of Holomorphic Functions*, 1994. ISBN 978-88-7642-275-1

C. PETRONIO, *Standard Spines and 3-Manifolds*, 1995. ISBN 978-88-7642-256-0

I. DAMIANI, *Untwisted Affine Quantum Algebras: the Highest Coefficient of* det H_η *and the Center at Odd Roots of 1*, 1996. ISBN 978-88-7642-285-0

M. MANETTI, *Degenerations of Algebraic Surfaces and Applications to Moduli Problems*, 1996. ISBN 978-88-7642-277-5

F. CEI, *Search for Neutrinos from Stellar Gravitational Collapse with the MACRO Experiment at Gran Sasso*, 1996. ISBN 978-88-7642-284-3

A. SHLAPUNOV, *Green's Integrals and Their Applications to Elliptic Systems*, 1996. ISBN 978-88-7642-270-6

R. TAURASO, *Periodic Points for Expanding Maps and for Their Extensions*, 1996. ISBN 978-88-7642-271-3

Y. BOZZI, *A study on the activity-dependent expression of neurotrophic factors in the rat visual system*, 1997. ISBN 978-88-7642-272-0

M. L. CHIOFALO, *Screening effects in bipolaron theory and high-temperature superconductivity*, 1997. ISBN 978-88-7642-279-9

D. M. CARLUCCI, *On Spin Glass Theory Beyond Mean Field*, 1998. ISBN 978-88-7642-276-8

G. LENZI, *The MU-calculus and the Hierarchy Problem*, 1998. ISBN 978-88-7642-283-6

R. SCOGNAMILLO, *Principal G-bundles and abelian varieties: the Hitchin system*, 1998. ISBN 978-88-7642-281-2

G. ASCOLI, *Biochemical and spectroscopic characterization of CP20, a protein involved in synaptic plasticity mechanism*, 1998. ISBN 978-88-7642-273-7

F. PISTOLESI, *Evolution from BCS Superconductivity to Bose-Einstein Condensation and Infrared Behavior of the Bosonic Limit*, 1998. ISBN 978-88-7642-282-9

L. PILO, *Chern-Simons Field Theory and Invariants of 3-Manifolds*, 1999. ISBN 978-88-7642-278-2

P. ASCHIERI, *On the Geometry of Inhomogeneous Quantum Groups*, 1999. ISBN 978-88-7642-261-4

S. CONTI, *Ground state properties and excitation spectrum of correlated electron systems*, 1999. ISBN 978-88-7642-269-0

G. GAIFFI, *De Concini-Procesi models of arrangements and symmetric group actions*, 1999. ISBN 978-88-7642-289-8

N. DONATO, *Search for neutrino oscillations in a long baseline experiment at the Chooz nuclear reactors*, 1999. ISBN 978-88-7642-288-1

R. CHIRIVÌ, *LS algebras and Schubert varieties*, 2003. ISBN 978-88-7642-287-4

V. MAGNANI, *Elements of Geometric Measure Theory on Sub-Riemannian Groups*, 2003. ISBN 88-7642-152-1

F. M. ROSSI, *A Study on Nerve Growth Factor (NGF) Receptor Expression in the Rat Visual Cortex: Possible Sites and Mechanisms of NGF Action in Cortical Plasticity*, 2004. ISBN 978-88-7642-280-5

G. PINTACUDA, *NMR and NIR-CD of Lanthanide Complexes*, 2004. ISBN 88-7642-143-2

Fotocomposizione "CompoMat", Loc. Braccone, 02040 Configni (RI) Italia
Finito di stampare nel mese di marzo 2016
presso le Industrie Grafiche della Pacini Editore S.r.l.
Via A. Gherardesca, 56121 Ospedaletto, Pisa, Italia